Water Security in a New World

Series editor
Zafar Adeel, Pacific Water Research Centre, Simon Fraser University, Burnaby, BC, Canada

More information about this series at http://www.springer.com/series/13406

Zafar Adeel • Robert G. Wirsing
Editors

Imagining Industan

Overcoming Water Insecurity in the Indus Basin

Springer

UNITED NATIONS UNIVERSITY
UNU-INWEH
Institute for Water, Environment and Health

Editors
Zafar Adeel
United Nations University
Institute for Water, Environment and Health
Hamilton, ON, Canada

Robert G. Wirsing
Formerly, Georgetown University
School of Foreign Service in Qatar
Doha, Qatar

Water Security in a New World
ISBN 978-3-319-32843-0 ISBN 978-3-319-32845-4 (eBook)
DOI 10.1007/978-3-319-32845-4

Library of Congress Control Number: 2016951442

Printed on acid-free paper

This Springer imprint is published by Springer Nature
The registered company is Springer International Publishing AG Switzerland

Preface

The world is encountering an era in which new threats and opportunities abound. Terrorism threatens peace and harmony far and wide, but equally challenging are a host of threats stemming from what many consider as even more menacing developments such as overpopulation, climate change, loss of species diversity, failures of governance, proliferation of weapons of mass destruction, spread of epidemic disease, and mounting resource scarcities, to name only some of them. Whatever their disciplinary leanings, practically all serious observers of world affairs characterize the present era as one unusually beset with a myriad of deeply disturbing trends and, as well, with more than the usual turbulence and unpredictability. Some of these observers determinedly busy themselves nevertheless trying to tease from the chaos at least the broad outlines of the future, and they do not infrequently offer up alarming and conflict-laden scenarios under the ominous label of the "coming global disorder."

It is thus conceivable that the subject of this book might seem imprudently optimistic, especially since it is about a region of the globe that fares dismally in virtually every assessment of the world's danger spots. When it comes to insecurity of any kind, the Indus Basin is not perceived as a bastion of peace, harmony, and well-being of the populace. And yet it was not really optimism that drove the *Imagining Industan* project to completion but a combination of the available opportunities and what might happen if they are not availed. There was a significant consensus among all the book's contributors that further delay in tackling collectively the region's widely shared and massive problem of water insecurity probably risked intensifying already considerable tensions among the four states sharing the basin (Afghanistan, China, India, and Pakistan). Delay most certainly also posed a huge risk to the economic potential and well-being of the many millions who now inhabit the basin as well as of those who come after.

This is therefore a book which takes a deliberately pragmatic and sober look at the way things are today with respect to utilization of water resources in the Indus Basin. Without excessive expectations (but with an abundance of concern for the region's future water security), it examines the reasons for the meager level of cooperation one finds among the basin's four riparian neighbors, the costs of noncooperation among them in regard to these resources, and, finally, some of the practical things that are being done now or can be done in the near future to move these states, however slowly and awkwardly, in the direction of cooperative, integrated management of the basin's water resources. That objective is obviously not one that will be easily achieved. However, movement in its direction is achievable, and, as many of the chapters in this book testify, the benefits such movement will bring are almost certainly immense.

Through their superb contributions and collective wisdom, the authors have undertaken a skilled and thorough assessment of the problem of water insecurity in the Indus Basin and explored some new concepts and innovative, unconventional solutions. We believe that this exploration provides the requisite evidence that can be taken up by the governments and other stakeholders when they consider the future of the Indus Basin.

Hamilton, ON, Canada — Zafar Adeel
Doha, Qatar — Robert G. Wirsing
April 2016

Acknowledgments

Support for this book's development came from many sources and took many forms. Of maximum importance, naturally, was the superb cooperation of the contributing authors. They came from many parts of the world to participate in an authors' workshop, held in Hamilton, Ontario, from 16 to 18 September 2015. Hosted by the United Nations University's Institute for Water, Environment and Health (UNU-INWEH), the workshop supplied a needed environment at an early point in the project for spirited and unfettered critical examination of draft papers. The authors in succeeding months accepted the editors' requests for revision without complaint and worked hard to meet deadlines.

The editors are indebted to the UNU-INWEH for financial support of various sorts at critical stages of the project. They wish to express heartfelt gratitude to the staff of UNU-INWEH, in particular to Ms. Ann Caswell and Ms. Imelda Pangantihon, for indispensable administrative support at all stages of the project. Kristina Roic and Kerry O'Neill contributed their truly admirable skill at compiling summaries of workshop discussions.

For the two original and meticulously prepared maps of the Indus Basin in the introductory Chap. 1, the editors owe their sincerest thanks to Robert Laws, Digital and Media Services librarian at Georgetown University's School of Foreign Service in Qatar.

Springer Publishing Company deserves special thanks for early on recognizing the project's merit and for offering encouragement all the way to publication. In particular, the editors wish to express their deep gratitude to Magesh Kaarthick Sundaramoorthy, project coordinator (Books) for Springer, SPi Global, not only for the commendable care he took in overseeing preparation of the manuscript and monitoring the production process but also for the great patience he displayed as the editors applied final touches to the book in its last stages.

Lastly, we want to acknowledge the vast and immeasurable contributions to this project made by the necessarily unnamed but countless individuals who over the years through their writings informed the contributors to this volume about the Indus Basin. In no small sense, they are virtual and truly valued co-contributors.

Contents

About the Authors

Zafar Adeel serves as the director of United Nations University's Institute for Water, Environment and Health, with responsibilities for research leadership and management of this United Nations think tank since 2006. He has extensive work experience in a broad range of international development and policy issues. His research focus is on water security and the 2030 Agenda for Sustainable Development, with particular attention to the nexus of water, food, and energy security. He has led the development of a south-south network of scientists working in water-scarce countries, particularly those in Africa, Middle East, and Asia; through his editorial lead, this network has published eight books in the UNU Desertification Series. He presently serves as the series editor for a book series by Springer: "Water Security in a New World." He has studied formulation of environmental policy and governance at several levels; a key output on this topic was a book entitled *East Asian Perspectives in Environmental Governance: Response in a Rapidly Developing Region* (UNU Press 2003). He also teaches environmental policy courses and supervises graduate students in his role as an adjunct professor of engineering at the McMaster University.

Majed Akhter is assistant professor of geography at Indiana University Bloomington. His research interests include the politics of water development, water security, drone war and imperialism, infrastructures and regionalism, Marxist geographical theory, and the political and historical geography of Pakistan and South Asia. Majed's articles have been published in journals such as *Antipode*, *Political Geography*, *Geoforum*, and *Critical Asian Studies*. He also coauthored a policy piece on water. His current research examines the implications of massive Chinese infrastructural investments in the form of the "New Silk Road" for the political geography of state power and the politics of development in Pakistan.

Saleem H. Ali holds the chair in sustainable resource development at the University of Queensland, Australia, and is also an adjunct professor of environmental studies at the University of Vermont (USA), where he founded the Institute for Environmental Diplomacy and Security. His research focuses on the causes and consequences of natural resource conflicts and how ecological factors can promote peace. His awards include a Young Global Leader citation by the World Economic Forum and an Emerging Explorer award from the National Geographic Society. Professor Ali received his doctorate from MIT, his masters in environmental studies from Yale University, and his bachelor in chemistry (summa cum laude) from Tufts University.

Dustin Garrick is an assistant professor and Philomathia chair of water policy at McMaster University. He specializes in water allocation policy and river basin governance in water-stressed regions of Western North America and Southeast Australia. His work has been published in *Science*, *Annual Review of Environment and Resources*, *Philosophical Transactions of the Royal Society*, and several public policy and economics journals. He holds grants from the Canadian and Australian research councils to investigate adaptation to climate extremes and water allocation reform in North America and Australia.

He has a special interest in interdisciplinary and science-practitioner networks that connect science and policymaking. He currently serves on the management committee of the Food, Energy, Environment and Water (FE2W) Network. He recently served on the Global Water Partnership/OECD Task Force on Water Security and Sustainable Growth and is active on a number of international and comparative water policy projects.

Prior to joining McMaster in January 2014, Dr. Garrick was a research fellow at the University of Oxford (2011–2013) and a Fulbright scholar (2010–2011) in Australia, where he remains a research associate of the Centre for Water Economics, Environment and Policy at the Australian National University. In the course of this book project, he worked closely with practitioners at the Bureau of Reclamation (Colorado River), in the Columbia Basin Water Transactions Program, and throughout Australia's Murray-Darling Basin.

Douglas Hill is a senior lecturer in development studies at the University of Otago, New Zealand. His recent work examines the politics of transboundary water sharing in South Asia, with a particular focus on the role of civil society in multi-stakeholder dialogues. He has worked or held visiting positions at universities, research institutes, and think tanks in Australia, Germany, the Netherlands, and the United Kingdom.

Srikanth Kondapalli is professor in Chinese studies at Jawaharlal Nehru University. He is chairman of the Centre for East Asian Studies, SIS, JNU, thrice from 2008 to 2010, 2012 to 2014, and 2016 to 2018. He is educated in Chinese studies in India and China with a Ph.D. in Chinese studies. He learnt Chinese language at the Beijing Language and Culture University and was a postdoctoral visiting fellow at People's University, Beijing, from 1996 to 1998. He was a visiting professor at National Chengchi University, Taipei, in 2004; a visiting fellow at the China Institute of Contemporary International Relations, Beijing, in May 2007; an honorary professor at Shandong University, Jinan, in 2009, 2011, 2013, and 2015 and at Jilin University, Changchun, in 2014; a nonresident senior fellow at People's University since 2014;

and a fellow at Salzburg Global Seminar in 2010. He wrote two books (*China's Military: The PLA in Transition* in 1999 and *China's Naval Power* in 2001) and two monographs and coedited three volumes (*Asian Security* and *China* in 2004, *China and Its Neighbours* in 2010, and *China's Military and India* in 2012) and a number of articles in journals and edited volumes – all on China. He received the *K. Subrahmanyam Award* in 2010 for *Excellence in Research in Strategic and Security Studies*.

Bjørn-Oliver Magsig is a research fellow at the Helmholtz Centre for Environmental Research (Leipzig, Germany), holds a lectureship in international environmental law at Leipzig University, and is a Royal Society of Edinburgh European visiting research fellow at the IHP-HELP Centre for Water Law, Policy and Science (University of Dundee, Scotland). In his research, Bjørn-Oliver focuses on international environmental law, water diplomacy, and the links between natural resources, international security, and equity. He has extensive project experience in international law governing transboundary freshwater resources, climate change, and the marine environment, serves on the Managing Board of the European Environmental Law Forum (EELF), and is a member of the IUCN World Commission on Environmental Law.

David Michel is a nonresident fellow at the Stimson Center in Washington, DC, and a global fellow at the Geneva Centre for Security Policy. He was previously senior associate and director of the Environmental Security Program at Stimson. His work explores emerging governance challenges and security risks posed by global environmental change, focusing particularly on the Indo-Pacific region. Michel has written widely on transboundary water management, maritime resources policy, the international impacts and implications of global warming, and the possibilities for collective institutions to address common environmental problems. He has advised the National Intelligence Council and the US Departments of Defense, Energy, and State on water and climate policy issues. He frequently lectures on environmental security at universities in the USA and abroad and has consulted with NGOs participating in the global climate negotiation process. Michel earlier served as senior associate with the Center for Transatlantic Relations at the Johns Hopkins University. He was educated at Yale University, the École des Hautes Études en Sciences Sociales in Paris, and the Johns Hopkins University's School of Advanced International Studies.

Paula R. Newberg has been clinical professor of government and fellow of the Wilson Chair at the University of Texas at Austin since January 2013. Prior to coming to UT-Austin, she was the Marshall B. Coyne director of the Institute for the Study of Diplomacy at Georgetown University. A scholar and practitioner with wide-ranging experience in multilateral, nongovernmental, and nonprofit organizations, Dr. Newberg served as special advisor to the United Nations in Asia, Africa, Europe, and the Middle East from 1996 through 2005. She was previously a senior associate at the Carnegie Endowment for International Peace and a guest scholar at the Brookings Institution and taught in the graduate faculties at Columbia and Johns Hopkins Universities. Dr. Newberg's work focuses on the intersections of governance and foreign policy in crisis and transition states. She has written extensively on constitutional development and jurisprudence in Pakistan (*Judging the State*, Cambridge University Press), on the politics of assistance in and to conflict and post-conflict states (*Politics at the Heart*, Carnegie Endowment), on rights in conditions of insurgency (*Double Betrayal*, Carnegie Endowment), and on international politics as a contributing columnist for the *Los Angeles Times*, *The Globe and Mail*, and *YaleGlobal Online*.

Manzoor Qadir serves as assistant director of the Water and Human Development Program at United Nations University's Institute for Water, Environment and Health. His work is focused on biophysical and policy aspects of water recycling and safe and productive reuse, water quality and environmental health, and salt-induced land degradation. He has implemented multidisciplinary projects and directed research teams in Central Asia and Middle East and North Africa regions. He previously held professional positions as senior scientist jointly appointed by the International Center for Agricultural Research in the Dry Areas and the International Water Management Institute; visiting professor at the Justus-Liebig University, Giessen, Germany; and associate professor at the University of Agriculture, Faisalabad, Pakistan. He coedited a book entitled *Wastewater: Economic Asset in an Urbanizing World*, published by Springer in 2015. He is a fellow of the Alexander von Humboldt Foundation and serves on the editorial boards of several international journals.

Kristina Roic graduated with a master of arts in globalization studies from McMaster University and a graduate diploma from the United Nations University's Institute for Water, Environment and Health in 2015. Her research work focused on water conflict and cooperation in India and Pakistan, under the supervision of Dustin Garrick, associate professor at McMaster University and Philomathia chair of water policy, and Manzoor Qadir, assistant director of the Water and Human Development Program at the United Nations University's Institute for Water, Environment and Health. Kristina works as a speechwriter for Global Affairs Canada in Ottawa, Canada, where she has been writing speeches for the Minister of International Development and *La Francophonie* since 2009.

Ashok Swain is professor of peace and conflict research and the director of the Research School of International Water Cooperation at Uppsala University, Sweden. He received his Ph.D. from the Jawaharlal Nehru University, New Delhi, in 1991, and since then he has been teaching at Uppsala University. He has been a MacArthur fellow at the University of Chicago; visiting fellow at the UN Research Institute for Social Development, Geneva; and visiting professor at the University of Witwatersrand, the University of Science in Malaysia, the University of British Columbia, the University of Maryland, Stanford University, McGill University, and

Tufts University. He has written extensively on new security challenges and democratic development issues. His recent publications include *Managing Water Conflict: Asia, Africa and the Middle East* (Routledge 2004), *Struggle Against the State: Social Network and Protest Mobilization in India* (Ashgate,2010), *Understanding Emerging Security Challenges: Threats and Opportunities* (Routledge 2012), *Transboundary Water Management and the Climate Change Debate* (Routledge 2015), and *Emerging Security Threats in the Middle East: The Impact of Climate Change and Globalization* (Rowman and Littlefield 2016). He has also edited/coedited several volumes, including *Education as Social Action* (Palgrave Macmillan 2005); *Islam and Violent Separatism* (Kegan Paul 2007); *Globalization and Challenges to Building Peace* (Anthem 2007); *The Democratization Project: Opportunities and Challenges* (Anthem 2009); *The Security-Development Nexus: Peace, Conflict and Development* (Anthem 2013); and *Water Security*, 4 volumes (Sage 2014).

Robert G. Wirsing was professor of government at Georgetown University's School of Foreign Service in Qatar (SFS-Q) until his retirement at the end of 2015. Earlier he was a member of the faculty of the Asia-Pacific Center for Security Studies, Honolulu, Hawaii (2000–2008), and of the Department of Government and International Studies, University of South Carolina (1971–2000). A specialist on South Asian politics and international relations, he has made over 40 research trips to the South Asian region since 1965. His publications include *Pakistan's Security Under Zia, 1977–1988* (St. Martin's Press 1991); *India, Pakistan, and the Kashmir Dispute* (St. Martin's Press 1994); *Kashmir in the Shadow of War* (M. E. Sharpe 2002); *Religious Radicalism* and *Security in South Asia*, as coeditor (Asia-Pacific Center for Security Studies 2004); *International Conflict over Water Resources in Himalayan Asia*, as coauthor (Palgrave/Macmillan 2013); and *Bengal's Beleaguered Borders: Is There a Fix for the Indian Subcontinent's Transboundary Problems*, as coauthor (The Asia Papers, Center for International and Regional Studies, Georgetown University School of Foreign Service in Qatar 2016). His most recent research focuses primarily on the politics and diplomacy of water resource conflict in South Asia.

Asim Zia is currently serving as an associate professor of public policy and decision analysis in the Department of Community Development and Applied Economics, with a secondary appointment in the Department of Computer Science, at the University of Vermont. He is associate state director of Vermont EPSCOR and the director of the Institute for Environmental Diplomacy and Security (IEDS) at the University of Vermont. Dr. Zia is undertaking NSF-funded work on developing computational integrated assessment models that project and quantify high-resolution impacts of global climate change and land-use land cover change on watershed-scale hydrological systems and lake systems and their feedbacks on social systems. Dr. Zia has published widely on international climate policy, management of social ecological systems, risk communication, risk management, international development, tropical forest conservation, and adaptation to climate change in the US, Latin American, African, and Asian contexts. He is also known for his work on mainstreaming early warning systems in development and planning processes and designing institutions and governance networks to assess and mitigate risks from environmental, social, and political drivers of change. Asim Zia has published 40 peer-reviewed articles/book chapters and three books. Most recently, Rutledge Press published his book *Post-Kyoto Climate Governance: Confronting the Politics of Scale, Ideology and Knowledge*. He has served as a principal investigator, co-principal investigator, or co-investigator on 16 research grants worth more than $35 million. Grant funds have been secured from the National Science Foundation, MacArthur Foundation, US-DOT, US-DoD, and USDA. He has a Ph.D. in public policy from the Georgia Institute of Technology.

Part I
Introductory Perspectives

Chapter 1
Introduction

Zafar Adeel and Robert G. Wirsing

Abstract This chapter outlines the volume's overall content and objectives. It defines the problem of water insecurity in the Indus basin and provides essential background data on the basin's history, geography, demography, and hydrology. It offers an explanation for the enormous importance of the basin's water resources to stakeholders at both the state (Afghanistan, China, India, and Pakistan) and sub-state levels, clarifies the relationship between the Kashmir dispute and the waters of the Indus system, supplies an overview of the Indus Waters Treaty, and highlights the geopolitical significance to the international community of the Indus basin's water resource circumstances. It briefly identifies each chapter's main focus.

Keywords Indus basin • Water security • Water scarcity • Water conflict • Water cooperation

1.1 What This Book Is About

This book is about transboundary cooperation in regard to the water resources of the Indus basin. It includes consideration of the impediments to cooperation as well as of cooperation's future prospects. It assesses the probable benefits of cooperation to the Indus river system's riparian states along with the costs of noncooperation; and it identifies some of the ways in which cooperation can most fruitfully be pursued. In theoretical terms, it is about the potential for moving the basin's four co-riparian states (Afghanistan, China, India, Pakistan) *from* a rigidly statist (unilateral, sovereignty-based, or resource-dividing) model of water resource management and utilization *to* a flexibly basin-wide (bilateral or multilateral, integrated, or resource-sharing) model.

Z. Adeel
United Nations University, Institute for Water, Environment and Health, Hamilton, ON, Canada
e-mail: adeel_unu@yahoo.com

R.G. Wirsing (✉)
Formerly, Georgetown University School of Foreign Service in Qatar, Doha, Qatar
e-mail: robertgw36@gmail.com

Z. Adeel, R.G. Wirsing (eds.), *Imagining Industan*, Water Security in a New World, DOI 10.1007/978-3-319-32845-4_1

The Indus basin lies in a part of the world where intense distrust, chronic conflict, and bitterly contentious water policies have a long history; and thus the book's authors harbor no illusions about the ease of accomplishing the admittedly herculean tasks implicit in fostering transboundary cooperation. Nevertheless, the book is premised on the conviction that the basin's inhabitants would benefit greatly from a fundamental reconfiguration of its water resource management and utilization. This reconfigured basin – termed Industan for the sake of this book – would be one in which the capacity for enhancing water security for all the basin's occupants would grow at a pace at least matching that of the mounting threat of water scarcity. In short, the discourse around Industan is thus guided by the belief that transboundary cooperation is an urgently required component of the region's toolkit for combating the steadily worsening problem of water insecurity in the Indus basin.

The volume's title is a play on words and is not meant to imply that changes to existing political boundaries are either required or even desirable; it most certainly should not be understood as headlining an outrageously unrealistic argument for the undoing of the 1947 Partition or fashioning an entirely new country. What the volume does envision – and labors to encourage – is a definite departure from business as usual responses to the Indus basin's vast and multifaceted emerging freshwater crisis. Such responses, it is becoming increasingly clear, are an invitation to eventual catastrophe.

The Industan this volume imagines (see Fig. 1.1) clearly cannot be realized without fundamental changes of some sort being made in the Indus basin's political dynamics, specifically in the way the riparian states and sub-state entities manage their water resource relationships with one another. Even in its infant stages, it would differ substantially from what is presently an area whose heavily river-dependent co-riparian neighbors – of both *inter*state and *intra*state kinds – are more often seen at loggerheads over water resources than endeavoring to cooperate.

The prospects this volume investigates for reconfiguration of the basin draw its authors into highly diverse topics. These naturally include consideration of the potential benefits to be gained from revisiting (replacing or revising) the 1960 Indus Waters Treaty, whose explicit admonition to its co-signatories to engage after its signing in cooperative water development activities has been steadfastly ignored for over half a century, but they also include examination of the potential for serious cooperation in, for example: combating climate change and the threat of glacial melt, the development of hydroelectric power, the deterrence of pollution and improvement of water quality, heightened deployment in the agricultural domain of water saving and demand lessening technologies, the sharing of data about flooding and surface/subsurface water depletion, the search for new techniques for heightening water supply, and the introduction of novel institutions for water management at local, provincial, national, and international levels.

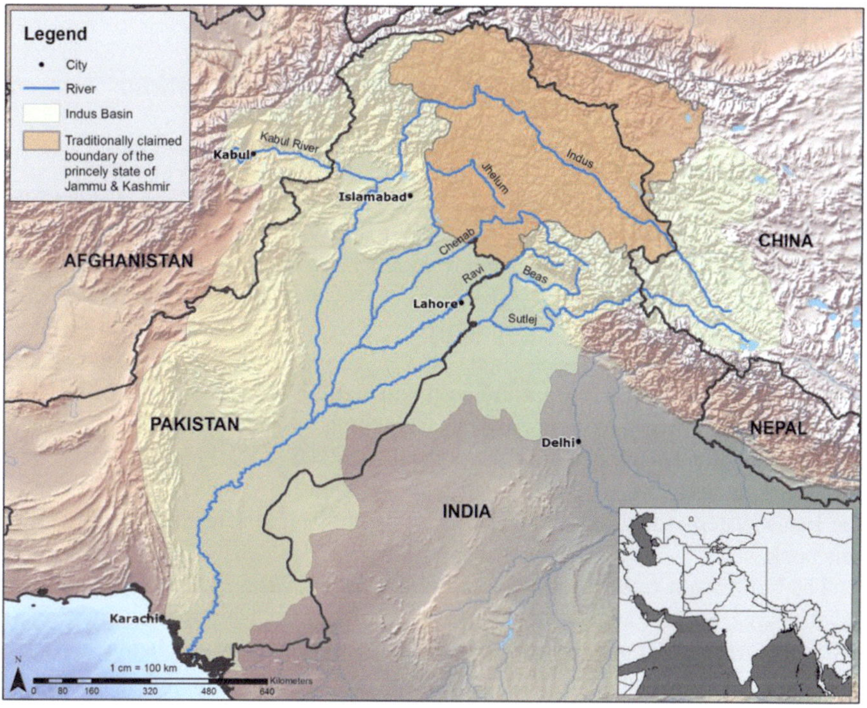

Fig. 1.1 The geographical extent of a hypothetical Industan

The volume also reviews ways in which the two hydrologically lesser (but by no means unimportant) Indus basin co-riparian states – namely, Afghanistan and China – can be successfully incorporated into the transboundary water cooperation equation. This is an increasingly urgent matter, particularly as Afghanistan's developmental agenda imposes new demands on the already over-extracted Indus waters and also as China's own freshwater crisis continues to heighten Beijing's interest in nearby Tibetan water resources. The volume does not neglect the interconnectedness of the Indus basin with the wider world: the continuous two-way transactions between the basin and the changing regional and global political order carry enormous implications both for water cooperation in the region and for regional and global security.

The ultimate objective of this volume is to kindle serious discussion of the transboundary cooperation that is needed to confront what more and more water experts believe is developing into one of the planet's most gravely threatened river basins. It is thus both assessment of the current state of play in regard to water security in the Indus basin and recommendation about where to go from here.

1.2 What the Indus Basin Encompasses

1.2.1 Geography of the Region and Distribution of Water Resources

The total area of the basin is approximately 1.12 million km^2, ranking it about 21st in size among all the world's river drainage or catchment areas. It embraces a whopping 65 % of Pakistan's total area, with what remains of the basin divided disproportionately among India (14 %), Afghanistan (11 %), and China (1 %). As for each country's share of the basin's territorial extent, the division is only a bit more even, with Pakistan commanding 47 %, India 39 %, China 8 %, and Afghanistan 6 %. The total water withdrawal in the basin is estimated at 299 km^3, irrigation withdrawal amounting to 93 % of that, with Pakistan accounting for about 63 %, India for 36 %, Afghanistan for 1 %, and China for a meager 0.04 % (AQUASTAT 2011). The evident asymmetry we see here – and its implications for each country's stakes in the Indus system – naturally contributes to the political puzzles later dealt with in this book.

The Indus mainstream, ranking 22nd in length (3180 kms/1976 miles) and 17th in discharge, stands among the world's greatest rivers. Originating in China's Tibetan Plateau, it is the axial branch of the Indus system of rivers that includes the flow of six major tributaries – one of them (the Sutlej), like the Indus itself, arising in China's Tibetan Plateau, four of them (Ravi, Beas, Jhelum, and Chenab) arising in India, and a sixth (the Kabul) arising in Afghanistan. Three of these rivers (Ravi, Beas, and Sutlej) are formally designated in the bilateral (India and Pakistan) 1960 Indus Waters Treaty as the Eastern rivers (their waters belonging to India), while the Jhelum and Chenab, along with the Indus mainstream, are designated as the Western rivers (belonging to Pakistan).

1.2.2 Demographic Characteristics

"Very roughly," according to UN Food and Agriculture Organization AQUASTAT data, "at least 300 million people are estimated to live in the Indus basin" (AQUASTAT 2011). The largest portion of this number (certainly over half), owing both to its disproportionately larger share of the basin's territorial extent and also to the fact that nearly all of its major cities lie within the basin, belongs to Pakistan; but Afghanistan and especially India can also claim fairly hefty shares. Judging from population growth figures, Afghanistan and Pakistan in particular are bound to increase their shares of the basin's population quite substantially in the coming decades. This is visible both in the two countries' cumulative fertility rates (average number of children per woman) in the most recent half decade 2010–2015, which show Afghanistan (5.13) and Pakistan (3.72) well ahead of India (2.48) and China (1.55), and also in cumulative annual population percentage increases for the same

5 years, again showing Afghanistan (70.9) and Pakistan (37.4) well out in front of China (27.4) and India (19.1) (United Nations Department of Economic and Social Affairs 2015). Put in terms of plain population numbers, the Indus basin grows steadily in importance to its two westernmost neighbors.

No less apparent, given the confessional identities dominant now in these four countries, is that the basin's population will tilt just as steadily in coming decades toward increasing Muslim dominance. This trend is perfectly consistent with global trends. According to recent Pew Research Center calculations, the world's Muslims, now growing in number nearly twice as fast as the overall global population (1.8 % as compared with 1.1 %), are projected to grow from 23.2 % of world population in 2010 to 29.7 % in 2050. By the latter date, three of the four basin co-riparian countries are expected to rank first, second, and tenth among the world's largest Muslim populations – India, with 310.66 million Muslims (11.2 % of the total world Muslim population); Pakistan, with 273.11 Muslims (9.9 %), and Afghanistan, with 72.19 Muslims (2.6 %) (Pew Research Center 2015). Interestingly, if one lumps together the projected Muslim populations in 2050 of these three countries with the South Asian region's other Muslim-majority country, Bangladesh, the total – 838.32 million – is slightly higher than the number of Indians who returned themselves as followers of the Hindu religion in India's 2001 census (827 million, 80.5 % of the total). These figures, one hardly needs to add, teem with political as well as cultural implications. Of course, the basin's overall religious heterogeneity will continue to prevail, with large numbers of Sikhs, Buddhists, and Christians sprinkled among the Muslims and Hindus. Heterogeneity will continue to prevail also ethno-linguistically, with strong representation in the basin population of Urdu, Punjabi, Sindhi, Pashto, Dari, Hindi, and Kashmiri speakers.

1.3 Why the Water Resources of the Indus Basin Are Important to the Co-riparian States

The waters of the Indus basin are indeed the lifeblood of the basin's inhabitants. The basin provides water for agricultural irrigation and industrial requirements, hydroelectric power and flood control, recharging of aquifers, and maintenance of wetlands and coastal zones, as well as for transportation, fisheries, and domestic human consumption. About the towering importance of the basin's waters in the lives of its inhabitants, there is now a huge and growing literature.[1] And much will be said

[1]Among the more useful of many recent studies on the Indus basin's water resources (and especially on Pakistan's water crisis) are Indus Basin Working Group, *Connecting the Drops*: *An Indus Basin Roadmap for Cross-Border Water Research, Data Sharing, and Policy Coordination* (Washington, DC: The Stimson Center, in collaboration with the Observer Research Foundation and the Sustainable Development Policy Institute, 2013), at http://www.stimson.org/research-pages/connecting-the-drops/, retrieved 16 October 2013; *Michael* Kugelman and Robert M. Hathaway (eds), *Running on Empty*: *Pakistan's Water Crisis* (Washington, DC: Woodrow Wilson International Center for Scholars, 2009), at https://www.wilsoncenter.org/sites/default/

about these things in the pages of this book. Here we elaborate on two of the more important ways in which the Indus waters impact the lives of the basin's inhabitants – agricultural irrigation and hydroelectric power.

1.3.1 Irrigation Waters

The Indus system of rivers, which includes numerous smaller tributaries draining from the Hindu Kush, Karakoram, and Himalayan ranges, houses today on Pakistan's side of the border the fifth largest irrigation system in the world (and the world's largest *contiguous* irrigation system), comprising what is called the Indus Basin Irrigation System (IBIS). This system consists of three large (1000 MW or more) hydropower dams (Mangla on the Jhelum, Tarbela on the Indus, and Ghazi Barotha also on the Indus), the first two of which, together with the Chashma Barrage on the lower Indus, are Pakistan's only major storage dams; about 50 smaller dams (and many more under construction); 23 or so barrages and head-works; and a mammoth network of 12 inter-river link canals, 45 irrigation canals, over 107,000 watercourses, and millions of small farm channels and field ditches – together comprising over a million miles of water conveyance (Aquastat 2011). Begun in 1849 during the British colonial era, the network the British built over the course of nearly a century before hauling down the Union Jack over British India was encumbered neither with the territorial division of the 1947 Partition nor with the riverine division that arose from the 1960 Indus Waters Treaty. What they built was an integrated basin-wide irrigation system, about 90 % of it located in what would later become Pakistan and the rest in Indian Punjab – both an engineering marvel and, potentially, a model for transboundary cooperative development in the post-independence period (see Fig. 1.2). Just such a model, grounded in the notion that the Indus system was a "natural" entity that needed to be retained as a "depoliticized space" following Partition, was famously – and unsuccessfully –urged upon India and Pakistan in 1951 by David Lilienthal, a prominent American water bureaucrat (Haines 2014).

The critical role irrigation plays in Indus basin agriculture – and in the food security of the people living in and around the basin – can hardly be exaggerated. By far the greatest concentration of the world's irrigated land – close to 65 % of it – is in Asia, about 35 % of it in South Asia. As can be seen in Table 1.1, the four

files/A, retrieved 29 August 2015; Daanish Mustafa, Majed Akhter, and Natalie Nasrullah, *Understanding Pakistan's Water-Security Nexus* (Washington, DC: United States Institute of Peace, 2013), at http://www.usip.org/sites/default/files/PW88_Understanding_, retrieved 29 August 2015; Gitanjali Bakshi and Sahiba Trivedi, *The Indus Equation* (Mumbai: Strategic Foresight Group, 2011), at http://www.strategicforesight.com/publication_pdf/10345110617.pdf, retrieved 29 August 2015; and Muhammed J. M. Cheema and Prakashkiran Pawar, *Bridging the Divide: Transboundary Science & Policy Interaction in the Indus Basin* (Washington, DC: The Stimson Center, March 2015, at http://www.stimson.org/images/uploads/research-pdf, retrieved 29 August 2015.

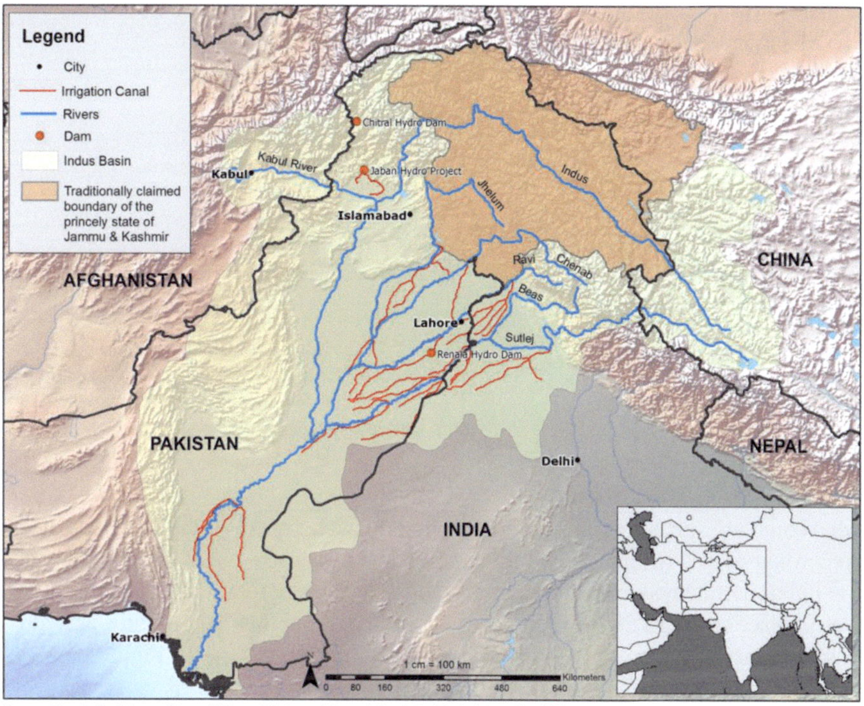

Source: Google Earth Layer Shows History of Water and Dams in the Indus Basin, International Rivers

Fig. 1.2 British-built Indus Basin irrigation canals 1947

Himalayan Asian states of China, India, Pakistan, and Bangladesh, ranking among the top ten irrigated countries in the world, together account for just over half (50.33 %) of the world's irrigated lands. China, India, and Pakistan alone account for four-fifths of the irrigated acreage in Asia. The highest irrigation density of all is found in Northern India and Pakistan and in the Ganges and Indus basins, and with its irrigated land at just under 70 % of its total land area under agricultural production, Pakistan ranks number one in the world in density of irrigation. More than 95 % of irrigation in Pakistan is located in the Indus basin, and agriculture contributes 21.4 % to the country's GDP and employs about 45 % of its labor force. In short, irrigated agriculture is hugely important in Asia, especially in South Asia, most especially in Pakistan.

1.3.2 Hydroelectric Power

The six countries sharing Asia's Hindu Kush Himalayan (HKH) range of mountains (Afghanistan, Bhutan, China, India, Nepal, and Pakistan) reportedly together have plans afoot to build more than 500 new hydroelectric dams in the HKH region in

Table 1.1 Top ten irrigated countries in the world (2003–2009)

Country	Area irrigated (million hectares)
India	60.85
China	57.78
USA	22.39
Pakistan	19.59
Iran	8.70
Mexico	6.32
Turkey	5.34
Thailand	4.99
Bangladesh	4.73
Indonesia	4.50

Source: International Commission on Irrigation & Drainage (ICID) Data Base; at: http://www.icid.org/imp_data.pdf, accessed 21 May 2011

what some are speaking of as the most massive "water grab" in the world's history. According to one recent study, India alone aims to construct 292 dams throughout the Indian Himalayas over the next couple of decades, a feat which, if carried out, would place dams in 28 of 32 major river valleys, doubling India's current hydropower capacity, and tagging the Indian Himalayas with "one of the highest average dam densities in the world, with one dam for every 32 km of river channel" (Grumbine and Pandit 2013; Vidal 2013). China, which already accounts for about 20 % of the world's large dams, has plans for about 100 dams in the HKH.

Signs of the "grab" are conspicuous in the Indus basin, where all four of the Indus riparian countries are energetically engaged in it. At the tail end of the dam-building frenzy is Afghanistan, which currently has a total of four major hydropower dams in the Indus basin, only one of them (Naghlu dam), at 100 MW installed capacity (and generally in disrepair), qualifying for that classification. India has recently completed the long-stalled construction of the Salma Dam in the western part of Afghanistan ('Thank You India' 2015), and there are reports that India plans to help Afghanistan build 12 dams on the Kabul River, a development signaling the rapidly tightening relationship between Afghanistan and India and causing considerable worry on the Pakistan side of the border (Kochhar 2015; Mustafa 2011).

When it comes to dam building, China too is yet a very small player in the Indus basin. Its construction of a small hydroelectric station on the Sutlej River in the western part of the Tibetan Autonomous Region was reported in the Indian media in June 2006 ("China Quietly Builds" 2006), and by 2010, it had completed a medium scale dam on the Indus river close to Demchok, Ladakh (Sering 2010). Both dams set off alarm bells among downstream users, especially of India, several of whose states are heavily reliant on the waters of the Sutlej. These Indian users are increasingly apprehensive that China, which supplies about half of India's transboundary freshwater, may soon "emerge as the ultimate controller of water for nearly 40 % of the world's population" (Vidal 2013).

As noted earlier, Pakistan has an abundance of small dams and barrages in the Indus basin and three major hydropower dams – Mangla on the Jhelum, Tarbela on the Indus, and Ghazi Barotha also on the Indus. A fourth major hydroelectric project (the Neelum/Jhelum Hydroelectric Project) is in the mid-stages of construction (with Chinese assistance) on the Jhelum River; and a fifth (the mammoth 4500 MW Diamer-Bhasha Hydroelectric Project), carrying an estimated price tag of US$13.683 billion, is in the initial stages of construction (also with Chinese assistance) on the upper Indus river. Pakistan, facing a demand for electrical power growing at 10 % annually and with hydropower accounting for a relatively meager 37 % of its energy mix, had an installed generating capacity in 2013 of 20,822 MW. According to the government's Water and Power Development Authority (WAPDA), Pakistan expects to raise this to 42,000 MW by 2020 (Pakistan WAPDA 2013). Estimates of its longer term hydroelectric potential run as high as 100,000 MW. Pakistan presently can store only 15 MAF (million acre feet) or around 13 % of the annual flow of its rivers – a capacity far below most other irrigation-dependent countries and even this figure is in grave jeopardy due to the rapid depletion of storage capacity in Pakistan's major dams due to sediment buildup (Pakistan WAPDA 2013).

India, like Pakistan, has an abundance of water resources, and it has relied fairly heavily on hydroelectric power. In recent years, it has ranked seventh globally – after China, Canada, Brazil, the USA, Russia, and Norway, in that order – in hydropower generation. It possesses an estimated 148,700 MW of installed capacity potential, placing it sixth globally (after China, Brazil, the USA, Canada, and Russia, in that order) in economically exploitable hydropower. Nevertheless, in spite of chronic and economically costly power outages, only about 20 % of India's hydropower potential has been developed thus far, and hydropower's share of the country's total energy mix stood (in 2010) at a relatively modest 24.7 % (Central Energy Administration 2010; Wirsing et al. 2013). This goes a long way to explain India's ambitious current plans for dam construction in the HKH mountain range pointed out above.

In the Indus basin, India presently has six commissioned large dams – two on the Sutlej (Bhakra, Nangal), two on the Beas (Pandoh, Pong), and two on the Chenab (Salal, Baglihar). Revealed by the Indian side during the arbitral proceedings conducted of the Baglihar Hydroelectric Project (2005–2007) by the Swiss Neutral Expert Raymond Lafitte was that the Baglihar was only 1 of 16 Indian hydropower projects (already built, under construction, or planned) on the Chenab River alone and that these were scarcely half of what India was planning on the upper waters of the Indus system in the state of Jammu and Kashmir. Indeed, India conceded at the time to having a total of 33 hydropower projects of varying sizes and in differing phases of planning or construction on various rivers in the state (Lafitte, *Expert Determination* 2007; Wirsing et al. 2013). Clear to all is that the Indus river system in Kashmir enjoys great prominence (and, in fact, is second only to the mountainous areas in India's northeast) in India's plans for resolving the country's power shortages by determined exploitation of the country's vast hydropower potential.

The importance of the Indus basin to the co-riparian states extends far beyond the immediately tangible material benefits its waters bring to the people who live there. This is because, failing cooperation, the future of this basin – the future water security of generations to come – remains forever in great jeopardy. Cooperation will bring not just more or cheaper electric power or more reliable water supply for the basin's farms; more than that, it holds the possibility for radical transformation in the lives of its inhabitants. As things stand, hovering over the basin is a dark cloud of hostility and distrust that impedes innovation, breeds pessimism, and shrivels opportunities. Every single dimension of human activity in this basin, whether economic, social, or political, is burdened with a huge handicap arising from the basin's sovereignty-based, resource-dividing, and relentless unilateralism. In other words, the importance of the Indus basin to the co-riparian states is less in what it manages to deliver in irrigation waters or electric power, for instance, than in the costs, the immeasurable but almost certainly huge and long-term costs, of noncooperation in its management. We must leave it to our authors to persuade readers of this; but nothing is more crucial to our understanding of the freshwater crisis of the Indus basin than recognition of the potentially mammoth benefits that co-riparian cooperation – in other words a truly resource sharing model of management – can bring.

1.4 The Kashmir Dispute and the Indus System

From the above discussion, it should be easy to grasp a fundamental truth of the Kashmir dispute – namely, its intimate relation to the Indus system waters. Indeed, no harm would be done to the truth were we to speak as often of the Indus dispute as of the Kashmir dispute: in many respects, they are one and the same.

The Partition process in the summer of 1947 was itself as often concerned with the problem of dividing between the two successor states of India and Pakistan the vast British-built irrigation system, located in the soon-to-be-divided Punjab, as it was with establishing a boundary between them on grounds of religious identity. Crucial to recall here is that the only known instance of gross impropriety – of violating the mandated impartiality that was to govern Sir Cyril Radcliffe's deliberations over where to draw the boundary between India and Pakistan – cropped up a day or so before Radcliffe's announcement of the boundary award on 16 August 1947. In 1989, Radcliffe's private secretary at the time of Partition, Christopher Beaumont, revealed that Britain's last Viceroy, Lord Louis Mountbatten, apparently goaded into action by the Indian Prime Minister Jawaharlal Nehru, had brought pressure on Radcliffe to alter the first draft in favor of India over lunch a day or so before the award was announced. Thanks to Beaumont, we now know with certainty that the award was changed to hand over to India a substantial tract of land – the Zira and Ferozepur sub-districts of Punjab's Ferozepur district. Both were Muslim majority and, thus, should ordinarily have gone to Pakistan. But in these sub-districts were important headworks controlling the flow of the Sutlej River, a major source of irrigation water for both Indian and Pakistani Punjab. And so it was New Delhi's apprehensions over water, it appears, that were responsible for the results of the secret meeting between Mountbatten and Radcliffe – what Beaumont said had brought "grave discredit on both men" (Kaushik 2015; "Partitioning India" 2007; Wirsing 1994).

Over the nearly 70 years since it began in 1947, the Kashmir dispute has become a massive tangle of issues, some of them leading on a number of occasions to open warfare, having to do with much more than just water. But that "the river runs through it," to borrow from the title of Norman Maclean's 1976 novella, has been equally true. Indeed, there is no little irony in the fact that the dissatisfaction of Kashmiri Muslims with their post-Partition political fate has been increasingly expressed in recent years not merely as a demand for self-determination but also as frustration with not having a meaningful role in management of the Indus river system originating in their state. They were not consulted, they say, when the Indus Waters Treaty was drawn up, and the partitioning of the rivers resulting from that treaty leaves them essentially voiceless in regard to how those rivers are now exploited. Commenting on the IWT not long ago, a chief minister of the state reportedly declared it to be "the biggest fraud with the people of Jammu and Kashmir" (Haider 2014). Kashmiri political leaders now are demanding a greater share of the electricity generated in the state also that the government return control of existing hydropower projects to the state.

We need hardly add that any movement in regard to revising or discarding the IWT will have to take account of the interests not only of the stakeholder states – India, Pakistan, and potentially Afghanistan and China as well – but also of the sub-state parties, of whom the Kashmiri Muslims are only one of many, with stakes in Kashmir of their own.

1.5 The Indus Waters Treaty

The Indus Waters Treaty (IWT) was finally signed in 1960 after nearly a decade of frustratingly contentious negotiations between India and Pakistan. Mediated by the World Bank, it is a lengthy treaty with detailed annexes covering a variety of contingencies. It stands today, some 56 years after its signing, widely, but not universally, judged a successful transboundary water resources agreement. Some of its success is due, no doubt, to the fact that it was given large-scale international support, in the form both of extraordinarily patient third-party mediation by the World Bank and, no less important, also of generous agreement-enabling funds. In contrast to some other river treaties, including the Ganges Water Treaty of 1996 between India and Bangladesh, the IWT benefits from the fact that it was drawn up as a permanent agreement without an end date or requirement of negotiated renewal, and it has never been amended or updated. Undoubtedly contributing to its longevity, too, is that its designers built into the treaty carefully crafted institutional machinery (including the Indus Waters Commission) and elaborate conflict-resolving mechanisms (including a three-step conflict-avoidance procedural arrangement running from "question" to "difference" to "dispute") to ensure the treaty's durability. Only twice have the more serious mechanisms – those arising from "difference" or "dispute" – been activated, once in 2005 in regard to India's Baglihar Hydroelectric Project on the Chenab River and once again in 2010 in regard to its Kishanganga Hydroelectric Project on the Jhelum River. These cases are treated in some detail in later chapters of this book.

For now, what is important to keep in mind is that the IWT, in spite of its evident successes, is today the focus of serious controversy. There are calls for revising it and, on occasion especially from the Indian side, even to scrap it entirely. Part of the problem with it arises from the fact that the IWT was a bilateral river-dividing, not a water-sharing, treaty. That unique design clearly paved the way to the treaty's signing, since India and Pakistan were apparently in no mood in the 1950s to enter upon any arrangement that required more than nominal cooperation. Unfortunately, in the decades since its signing, circumstances affecting the basin – the explosion in population, growing water scarcity, and climate change only the most conspicuous of them – have highlighted the treaty's inadequacies and moved some observers to call for revisiting it.

Revisiting the IWT is clearly implicit in a reimagined Industan, and later chapters will be doing just that. Any revisit, however, will be up against a host of problems. Among them is the likelihood that Indians and Pakistanis won't be on the same page, differing not only on the need for change but also on its preferred contents. Serious consideration would have to be given to whether the IWT should remain a bilateral treaty or be widened to include one or both of the other two riparian countries, also to the question of whether it might be wiser to abandon the notion of a comprehensive agreement and negotiate instead separate agreements on such things as water quality, groundwater exploitation, disaster resilience, environmental protection, and so on. Perfectly obvious is that the IWT, for all its virtues, is a can of worms – a can that many feel must be opened.

1.6 Why the Indus Basin Is Important to the World

Cooperation among the riparian states of the Indus basin in the common quest for overcoming water insecurity is every bit as important to the world community as it is to the basin's riparian states. We have at least six reasons for thinking so.

The first is that the Indus basin riparian states rank exceptionally high on world indicators of great power status. Two of the four (China and India) rank first and third among the three largest economies in the world – one of them (China) having overtaken the USA as the world's largest economy by the end of 2014 (Stiglitz 2014). According to the United Nations *World Population Prospects: The 2015 Revision*, three of the four riparian states (China, India, Pakistan) currently rank first (1.376 billion), second (1.311 billion), and sixth (188.9 million) among the top populations in the world, and one of them (India), presently holding 18 % of the world population, is now expected to replace China as the world's most populous country by 2022 (United Nations Department of Economic and Social Affairs 2015). By 2030, another (Pakistan) is expected, according to a 2011 study by the Pew Forum on Religion & Public Life, to surpass Indonesia and emerge as the world's most populous Muslim country, by then rising to about 256 million (Pew Research Center 2011). The armies of three of the four riparian states (China, India, Pakistan) rank among the ten largest in the world, and the same three are nuclear weapon-armed members of perhaps the most exclusive weapon fraternity ever created.

The second is that the probability of armed conflict among the riparian states, exacerbated by the expected increase in resource scarcity and intensified resource rivalry, is unquestionably among the highest in the world. This doesn't mean that water wars are soon bound to break out in the Indus basin, though some of the area's best scholars seem to think so.[2] However, it does mean that a decent allowance has to be made for water scarcity's inclusion on the list of things the Indus basin's riparian states quarrel about. And their quarrels have often enough ended in war. We may note in this regard that all four of the riparian states already have a history of armed conflict, whether direct or by proxy, with one or more of their riparian neighbors in the last half century or so (India with both Pakistan and China, Afghanistan with Pakistan). Likewise, all four of the riparian states have major active territorial disputes with one or more of their riparian neighbors (India with both China and Pakistan, Pakistan with both Afghanistan and India). Three of the four riparian states face serious and persistent secessionist threats (China in Tibet and Xinjiang, India in Kashmir and parts of its northeast, Pakistan in Baluchistan). And finally, the Indus basin (and especially Pakistan) has acquired great notoriety as the operational base nowadays for numerous, fractious, reckless, and in some cases powerful trans-state terrorist-minded organizations, including the Taliban and Al-Qaeda (Laub 2013). None of these spell unavoidable water war, but, in combination with one another, they do spell more than average likelihood of interstate tension and the production of conditions characteristically associated with the prelude to war.[3]

The third is that there is clearly considerable risk of political instability arising from water scarcity in at least one of the Indus basin riparian states (Pakistan). American intelligence agencies and many in the media have habitually labeled Pakistan a strategic "flashpoint" and the world's "most dangerous place." Naturally,

---

[2] The most notable among them is the Indian scholar Brahma Chellaney, whose two most recent books offer the most scholarly and compelling arguments to be found in the burgeoning literature of water wars advocacy. See his *Water: Asia's New Battleground* (Washington, DC: Georgetown University Press, 2011) and *Water, Peace, and War: Confronting the Global Water Crisis* (New York: Rowman & Littlefield, 2013), pp 1–57. For a recent and persuasive argument that defines a position in the ongoing water war debate separate from both the water war and water peace advocacy schools, see the essay by the Crawford School of Public Policy's Paula Hanasz, *Troubled Waters: India and the Hydropolitics of South Asia*, Fearless India Occasional Papers on India-Australia Relations, v. 4 (Melbourne: Australia India Institute, Winter 2014).

[3] An especially noteworthy recent book that attempts to define what the author calls "the logic of great power conflict" is Christopher Coker, *The Improbable War* (London: Hurst & Company, 2015). While Coker's focus is on the possibility of war between China and the USA, the book supplies an unusually perceptive assessment of the logic of war in general.

this doesn't sit well with Pakistanis, and, in the face of far more violent and no less dangerous conditions in more than a few other places around the world, it is probably misleading. Nevertheless, with many authoritative predictions hovering over Pakistan – and over India, too, we might add – suggesting that its rendezvous with acute water scarcity is just around the corner, the anticipation of imminent political instability does not appear all that farfetched.

Pakistan's especially bleak water statistics were highlighted in a *Deutsche Welle* interview in February 2015 with Michael Kugelman, South Asia expert at the Woodrow Wilson Center in Washington, DC, who offered up a breathtaking litany of Pakistan's water woes. Pakistan, he said, has the fourth highest rate of water use in the world; and its water intensity rate (the amount of water, in cubic meters, used per unit of gross domestic product) is the highest in the world. It is, he said, the third most water-stressed country in the world[4], and its per capita water availability per annum, estimated at 1017 m^3, puts it just a hair above the widely accepted and much-dreaded scarcity threshold of 1000 m^3 per capita per annum (Dominguez 2015). The potential impact of all this on Pakistan's ability to keep pace economically with the parallel massive increase in population we've already taken note of is ominous. No surprise, then, that a recent International Monetary Fund assessment is that water scarcity may threaten all aspects of the country's economy (Kochhar 2015). And it seems reasonable to suppose that any prolonged spell of economic woes would very likely breed political instability.

The fourth reason for thinking that water resource cooperation among the riparian states of the Indus basin is every bit as important to the world community as it is to the basin's riparian states is that the Indus basin is widely expected to be among the world's worst-affected from climate change. Drought, desertification, weakened and/or variable monsoon rains, weather turbulence, flooding, sea level rise, and glacial retreat/melt are all routinely cited as probable developments impacting the basin in the twenty-first century, and there can be little doubt of their potentially harmful collective economic and political consequences. If Pakistan, for instance, is already water scarcity-endangered as a result of the demography-driven widening gap between supply and demand, how much wider might the gap become – and how much might this exacerbate the country's economic and political woes – when the difficulties of climate change are added in?

Let us hasten to emphasize here that no scientific consensus has yet developed in regard to the precise effects of hydro-climatic change on the basin. This is especially the case when it comes to the matter of glacial retreat/melt. In its chapter "Hydrology and Glaciers in the Upper Indus Basin," a recent and technically sophisticated World Bank report on the impacts of climate risks on the Indus basin begins

[4]The most recent ranking we know about of the world's most water-stressed countries places Pakistan 23rd of 167 countries by 2040. This obvious discrepancy is not surprising, since studies differ substantially in terms of the climate models and socioeconomic data adopted by the researchers. For this latest estimate, see Andrew Maddocks, Robert S. Young, and Paul Reig, "Ranking the World's Most Water-Stressed Countries in 2040" (Washington, DC: World Resources Institute, 26 August 2015), at http://www.wri.org/blog/2015/08/ranking-world's-most-water-stressed-countries, retrieved 4 September 2015.

the discussion with a warning that "considerable speculation but little analysis exists concerning the importance of glaciers in the volume and timing of flow in the Indus River and its tributaries, as well as on the potential impact of climate change on these rivers" (Yu et al. 2013). Abundantly clear from the ensuing discussion, however, is that the reliance of the basin's principal riparian, Pakistan, on irrigated land for more than 90 % of agricultural production, in company with the fact that an estimated 18 % of the total flow of the Indus river from its mountain headwaters comes from glacier runoff, most of the rest being meltwater from the winter snowpack, vastly increases its potential vulnerability to climate change (Yu et al. 2013, p 71). Moreover, even if we accept the reassuring conclusion of a recent China-supported study of climatic and hydrological changes in Pakistan's Upper Indus Basin (UIB) that higher temperatures are very likely to be accompanied by increased river flow in the Indus and that "water availability is likely to be increased in the twenty-first century and this may sustain water demands," we have to take note also of the same report's judgment that the increased river flow "might create more flooding in the first half of the century" (Ali et al. 2015).

The fifth is that the region in which the Indus basin is located collectively represents enormous market size and business opportunities. This is implicit, of course, in the mention we made above of the global economic rankings of China and India. Trade and investment opportunities are huge in both. China, for instance, ranks second (behind Canada and before Mexico) and India ranks eleventh in total trade volume, imports and exports, among America's top 15 trading partners (US Census 2013).

The sixth (implicit in all the others) is that the spillover effects of all of the above on the rest of the world cannot be contained and are likely to be sizeable. The basin's occupants are just too large, too important, and too integral a part of world affairs for their water woes to be simply swept aside as a matter of little consequence. The water insecurity of the Indus basin is a world, not just a local, problem.

1.7 Definitional and Data Considerations

On World Water Day 2013, UN-Water, a conglomerate of about 17 affiliated water-related groups assembled by the United Nations University, proposed a common definition of water security, the raison d'etre of this book. Water security, an unavoidably vague and potentially troublesome concept, was defined in the published analytical brief "as the capacity of a population to safeguard sustainable access to adequate quantities of acceptable quality water for sustaining livelihoods, human well-being, and socio-economic development, for ensuring protection against water-borne pollution and water-related disasters, and for preserving ecosystems in a climate of peace and political stability" (UN-Water 2013). The document went on to say that achieving water security required "allocation among users to be fair, efficient, and transparent; that water to satisfy basic human needs is accessible to all at an affordable cost to the user; that water throughout the water cycle is collected and treated to prevent pollution and disease; and that fair, accessible, and effective mechanisms exist to manage or address disputes or conflicts that may arise. The concept operates at all levels, from individual, household, and community to local, subnational, national, regional, and international settings and takes into account the variability of water availability over time" (UN-Water 2013).

Other definitions of water security exist and more attempts to grapple with the concept will undoubtedly be made. For the purposes of this book (and without any intention to force the adoption of this particular definition), this definition seems to us a good place to start.

Having furnished in the pages above a fair amount of statistical data in regard to the Indus basin, it seems advisable to acknowledge that in regard to most of the topics already covered – and many more yet to come – there is plenty of data discrepancy and grounds for disagreement. This is due in part to the frequent thinness, obsolescence, or complete absence of data. It is due also to the fairly innocent (albeit not necessarily forgivable) mishandling and misreporting of data. But, to be candid, it is also a product of willful efforts by governments and others to conceal or misrepresent the facts. After all, in regard to a basin where the stakes are exceptionally high, conflict endemic among the riparian entities (of both the state and sub-state genres), and the trust deficit simply enormous, one should not expect data in regard to anything of importance to be put forth unpremeditated and wholly uncontaminated by political and security considerations. Hence, skepticism is a needed safeguard.

This book is organized into four parts. Part One: Introductory Perspectives includes this introductory Chap. 1 and also Chap. 2: De-centering Per Capita Water Supply: The Political Ecology of the Water Scarcity/Security Nexus in the Indus Basin. These two chapters supply the broad geographic, demographic, hydrological, economic, political, historical, and theoretical perspectives that are needed in laying the foundation for the rest of the book. Part Two: The Costs and Scale of Transboundary Conflict also contains two chapters. Chapter 3 examines the costs of noncooperation among inter- and intrastate stakeholders, and Chap. 4 analyzes the problem of measuring the frequency and intensity of conflict and cooperation over water resources in India-Pakistan relations since independence was gained in 1947.

Part Three: The Potential for Transboundary Cooperation has five chapters. Chapter 5 focuses on the role of the 1960 Indus Waters Treaty in fostering transboundary cooperation. It assesses the past record of the treaty and evaluates alternative modalities for improving prospects for equipping it with greater future resiliency. Chapters 6 and 7 both focus on environmental security in the Indus basin, looking in Chap. 6 at the potential for transboundary cooperation in coping with climate change and then in Chap. 7 at prospects for harnessing regional organizations to the task of transboundary data sharing and resilience scenario development. Chapters 8 and 9 examine the potential for expanding basin-wide water resource management in the Indus basin – looking first, in Chap. 8, at the potential for such expansion in India-Pakistan relations and then, in Chap. 9, at this potential in China's relations with its Himalayan neighbors, India and Pakistan.

Part Four concludes the book. Chapter 10– The Role of International Development in Reimagining the Indus Basin – examines the role of the international community, international and regional organizations in particular, in the ongoing quest for transboundary cooperation in the management of water resources in the Indus basin. Finally, in the concluding chapter, Chap. 11, the coeditors summarize what the book's contributing authors tell us about a reimagined Industan – about the problem and potential of transboundary cooperation in regard to the water resources of the Indus basin – and, completing the task set forth in Chap. 1, offer their thoughts on why major geopolitical developments in the world in coming decades, at both the regional (South Asian and Asian) and global levels, will play a highly influential role in determining whether or not the basin's four co-riparian states (Afghanistan, China, India, Pakistan) do or do not move to a basin-wide (bilateral or multilateral, integrated or resource-sharing) model of water resource management.

References

Ali S et al (2015) Twenty first century climatic and hydrological changes over upper Indus basin of Himalayan Region of Pakistan. Environ Res Lett 10. https://iopscience.iop.org/1748-9326/10/1/014007. Retrieved 31 Aug 2015

AQUASTAT (UN Food and Agriculture Organization Information System on Water and Agriculture) (2011) Indus Basin, Water Report 37. http://www.fao.org/nr/water/aquastat/basins/indus/index.stm. Retrieved 24 Sept 2015

Central Energy Administration, Ministry of Power, Government of India. http://www.powermin.nic.in/Indian_electricity_scenario/introduction.htm. Retrieved 16 Sept 2010

China Quietly Builds a Barrage on Sutlej. Hindustan Times. 30 June 2006. http://www.phayul.com/news/article.aspx?id=13078&t=1. Retrieved 26 Sept 2015

Dominguez G (2015) Water scarcity 'Is Pakistan's Worst Nightmare'. Deutsche Welle, 7 February 2015. http://www.dw.com/en/water-scarcity-is-pakistan's-worst-nightmare/a-18557432. Retrieved 29 Aug 2015

Grumbine SE, Pandit MK (2013) Threats from India's Himalaya Dams. Science 339:6, 4 January 2013. https://www.sciencemag.org/content/339'6115/36.su. Retrieved 25 Sept 2015

Haider M (2014) Pakistan's water woes should not be blamed on 'Bogeyman' India. Dawn, 28 August 2014. http://www.dawn.com/news/1128132/print/print. Retrieved 26 Sept 2015

Haines D (2014) (Inter)Nationalist rivers? Cooperative development in David Lilienthal's plan for the Indus basin, 1951. Water Hist 6:133–151

Kaushik RK (2015) How ferozepur became ours. Times of India, 19 April 2015. http://timesofindia.indiatimes.com/india/How-Ferozepur-became-ours/articleshow/469742. Retrieved 26 Sept 2015

Kochhar K (2015) Impact of Afghanistan dams on Pak economy. Front Post, 24 May 2013. https://www.facebook.com/thefrontierpost/posts/456990237709802. Retrieved 25 Sept 2015

Kochhar K et al (2015) Is the glass half empty or half full? Issues in managing water challenges and policy instruments, IMF Staff discussion note. International Monetary Fund, Washington, DC, June 2015, p 7. https://www.imf.org/external/pubs/ft/sdn/2015/sd. Retrieved 29 Aug 2015

Lafitte R (2007) Expert determination on points of difference referred by the Government of Pakistan under the provisions of the Indus Waters Treaty. World Bank, Lausanne, 12 February 2007. http://www.pakistan.gov.pk/ministries. Retrieved 26 Sept 2015

Laub Z (2013) Pakistan's new generation of terrorists, CFR Backgrounders. Council on Foreign Relations, Washington, DC, 18 November 2013. http://www.cfr.org/pakistan/pakistans-new-generation-terrorists/p15422. Retrieved 23 Sept 2015

Mustafa K (2011) India to help Afghanistan build 12 dams on Kabul River. News, 12 May 2011. http://thenews.com.pk/PrintEdition.aspx?ID=5933&Cat=13&dt. Retrieved 25 Sept 2015

Pakistan Water and Power Development Authority (WAPDA) (2013) Hydro potential in Pakistan. Islamabad, October 2013. http://www.wapda.gov.pk. Retrieved 26 Sept 2015

Partitioning India over Lunch. BBC News. 10 August 2007. http://newsvote.bbc.co.uk/mpapps/pagetools/print/news.bbc.co.uk/2/hi/south_asia/6926464. Retrieved 26 Sept 2015

Pew Research Center (2011) The future of the global Muslim population: projections for 2010-2030. Pew Forum on Religion and Public Life, Washington, DC, January 2011. http://www.pewforum.org/files/2011/01/FutureGlobal. Retrieved 23 Sept 2015

Pew Research Center (2015) The future of world religions: population growth projections, 2010–2050. Pew Forum on Religion and Public Life, Washington, DC, 2 April 2015. http://www.pewforum.org/2015/04/02/muslims/. Retrieved 25 Sept 2015

Sering SH (2010) China builds dam on Indus near Ladakh. J Def Stud 4(2), April 2010. http://www.idsa.in/jds/4_2_2010_ChinaBuildsDamonIndus. Retrieved 26 Sept 2015

'Thank You India', Say Delighted Afghans as Salma Dam Nears Completion. Zee News, 5 August 2015. http://zeenews.india.com/news/india/thank-you-india-say-delighted-afghans-as-salma-dam. Retrieved 25 Sept 2015

United Nations, Department of Economic & Social Affairs, Population Division (2015) World population prospects: the 2015 revision, key findings and advance tables, Working paper No. ESA/P/WP.241. http://esa.un.org/unpd/wpp/Publications/Files/Key_Findings. Retrieved 24 Sept 2015

UN-Water (2013) Water security & the global water agenda: a UN-Water analytical brief. United Nations University Institute for Water, Environment & Health, Hamilton. http://www.unwater.org/downloads/watersecurity-an. Retrieved 25 Sept 2015

US Census, Top trading partners – December 2013. https://www.census.gov/foreign-trade/statistics/hi. Retrieved 24 Sept 2015

Vidal J (2013) China and India 'Water Grab' dams put ecology of Himalayas in danger. The Guardian, 10 August 2013. http://www.theguardian.com/global-development/2013/aug/10/china-india-water-grab-dams. Retrieved 25 Sept 2015

Wirsing RG (1994) India, Pakistan, and the Kashmir dispute. St. Martin's Press, New York, pp 24–25

Wirsing RG, Stoll DC, Jasparro C (2013) International conflict over water resources in Himalayan Asia. Palgrave Macmillan, Houndmills, pp 85–88

Yu W et al (2013) Indus Basin of Pakistan: impacts of climate risks on water and agriculture. World Bank, Washington, DC, p 57. https://openknowledge.worldbank.org/handle/10986/f3834. Retrieved 31 Aug 2015

Chapter 2
The Political Ecology of the Water Scarcity/Security Nexus in the Indus Basin: Decentering Per Capita Water Supply

Majed Akhter

Abstract The connection between water scarcity and water security in the Indus Basin is often understood through attention to the decline of physical water supply per capita. But water insecurity at the individual and regional scales is as much about political and social structures as it is about the absolute (or physical) availability of a natural resource. Drawing on insights from the interdisciplinary tradition of political ecology, this chapter highlights the importance of examining the interaction of absolute and structural scarcity in the historically and geographically specific context of the Indus Basin in northwest South Asia (and especially Pakistan and India). This chapter first evaluates the Indus Basin's ability to meet the basic physiological needs of the human population in terms of absolute supplies of water. The next section argues that the fundamental sociopolitical structures that shape the structural scarcity of water in the Indus can be rooted in the agroecological transformations of the basin since the late nineteenth century. The final section analyzes Indus Basin water security in the context of climate change through attention to the interaction of absolute and structural water scarcity.

Keywords Structural scarcity • Water security • Political ecology • Green Revolution • Canal colonies

2.1 Introduction

The Indus River system (henceforth, "the Indus") is the most significant source surface water in Pakistan and northwest India. Although portions of the Indus Basin also reach into Afghanistan, China, and Nepal (see Fig. 2.1), the overwhelming majority of its waters are withdrawn to support livelihoods in Pakistan and (to a lesser degree) in India. Hundreds of millions of small farmers divert millions of acre-feet every year from the Indus in an effort to feed one of the most densely

M. Akhter (✉)
Department of Geography, Indiana University, Bloomington, IN, USA
e-mail: maakhter@indiana.edu

Z. Adeel, R.G. Wirsing (eds.), *Imagining Industan*, Water Security in a New World, DOI 10.1007/978-3-319-32845-4_2

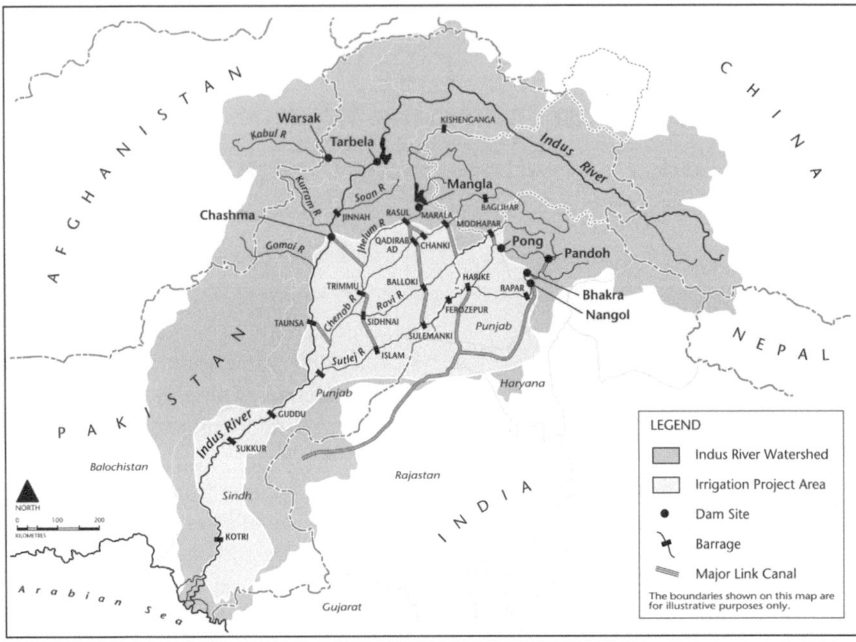

Fig. 2.1 The Indus Rivers (Image source: Mustafa et al. 2013)

populated swathes of land in the world. But rapid declines in per capita freshwater availability in the region over the past several decades and the specter of intense and unpredictable climate change have led many to fear that there simply isn't (or soon won't be) enough water to go around. Or, to put it a different way, some fear there are too many people around relative to the limited supply of natural resources.

This chapter begins from the premise that it is overly simplistic to understand physical scarcity, measured as per capita freshwater availability, as a major cause of water insecurity. I understand discourses that assert a direct causal relationship between physical scarcity and social crisis as part of the ideological mystification of the highly unequal social distribution of ecological goods (Mehta 2010; Harvey 1974). By ignoring the uneven power relations within society, these post-Malthusian narratives disproportionately blame marginalized social groups for increasing physical scarcity, even as these same groups are in fact the most vulnerable to rapid environmental change (Mehta 2010; Harvey 1974). Instead of attributing a causal efficacy to physical scarcity (too little water or too many people), a political ecological perspective understands historical change as process characterized by a complex and dynamic interaction between social and ecological forces.

This chapter examines the interaction between absolute (physical) and structural (social and historical) water scarcity in the Indus Basin. Although measurable decreases in water supply availability are undeniably important, we need a more nuanced understanding of the relationship between water scarcity and security. Natural resource scarcity is always and only experienced as interaction between absolute scarcity and structural scarcity – between ecological and social forces. This means that understanding variability of rainfall, glacier melt, and aquifer depletion should be complimented with an equally intense focus on factors such as history, technology, and political economy. In what follows, I first attempt to answer the basic question of whether there is enough water in the Indus Basin. An in-depth geographical analysis would situate human populations within a larger ecosystem of human and non-human connections - but for the purposes of critically engaging the question of water security and scarcity, I focus exclusively on human needs in what follows. The next section moves from an evaluation of absolute scarcity to a consideration of the structural aspects of water scarcity in the Indus. I sketch a hydro-ecological and political economic history of the Indus because it is precisely this history that provides the foundation for the production of structural scarcity in the region (Gilmartin 2015; Taylor 2014; Mustafa 2013). The two moments I focus on are British-led canal colonization of the late nineteenth and early twentieth centuries and the Green Revolution of the mid-twentieth century. My objective in this section is to show how the use and social distribution of water are shaped decisively by social, historical, and economic structures – not just the amount of total water available per capita. The next section reconsiders the question of water supply and security in the era of climate change through a prism that pays attention to the dynamic between absolute and structural scarcity.

2.2 Is There Enough Water in the Indus Basin?

There is an estimated long-term basin-wide surface water availability of 239–258 km^3 (194–209 million acre-feet or MAF) in the Indus Basin, of which roughly 179 km^3 (145 MAF) is extracted in Pakistan (Laghari et al. 2012, p. 1065). Surface water supply is supplemented by another important source. Below the rim of the Himalayas, the northern edge of the transborder Indo Gangetic plains, lies one of the largest freshwater aquifers in the world, spanning about 15 million acres of subsurface area. Replenishable groundwater resources in the Indus Basin in Pakistan are estimated at 63km^3 (51 MAF) (Laghari et al. 2012, p. 1065), while the Indian portion (approximated here as the total replenishable groundwater resources of the Indian state of Punjab) is 22.56 billion cubic meters (18 MAF) (Central Ground Water Board 2014, p. 43). Thus the total annual freshwater availability of the Indus Basin (not counting groundwater supplies outside of Pakistan and Indian Punjab) amounts to approximately 263–278 MAF.

Now that we have arrived at a necessarily rough estimate of the annual supply of Indus Basin surface and groundwater supplies, a consideration of aggregate demand is necessary to evaluate water scarcity and security. Although over 90 % of Indus waters that are currently withdrawn are used for purposes of irrigation, increased water demands in the future will likely come from cities and other centers of population growth for purposes of drinking water, cleaning, industrial uses, and sanitation. Population growth in Pakistan in particular has been very high, averaging an annual rate of 2.61 % in the period 1961–2011. Indeed, most writing on water scarcity and water policy in general in Pakistan and the Indus begins with the sobering fact that the per capita availability of water in Pakistan has decreased from 5260 km^3 in 1951 to roughly 1040 km^3 in 2010 (Government of Pakistan 2005). This is a decline in physical water supply/per capita of over 400 %. It seems self-evident then that "rapid population growth" will usher Pakistan into an "era of absolute water scarcity," (Briscoe and Qamar 2006, p. 3; Malik 2011). But the connection between declining water availability and social devastation in the Indus Basin, and especially Pakistan, is proclaimed more often than explained or defended. There is an unspoken assumption about the relationship between population, scarcity, and limited water supplies. It is important, however, to conceptually clarify and empirically verify how a calculated decline in per capita water availability ostensibly leads to crisis.

Population-based analyses tend to rely on a variety of Malthusian or post-Malthusian assumptions. The basic assumption is that population growth will outstrip the capacity of society to produce enough food for everyone. A limit, an ecological "carrying capacity" of a given region, will inevitably be reached, after which population must decrease for society to be sustainable. Pakistan, as will be recalled, is dependent on irrigated agriculture for virtually all of its food, and irrigation makes up the bulk of water extracted from the Indus River. Thus the argument about population-fueled water scarcity leading to national crisis depends on the following sequence: increased population rates result in decreased per capita water availability, and this results in decreased food production per capita, which could possibly result in starvation, economic collapse, and political conflict. An examination of the available data, however, shows that this sequence is not viable.

In fact, there is no clear relation between growing population rates, per capita water availability, and food production. Since the 1960s, growth in agricultural production has on average outpaced population growth. Let us return again to the case of Pakistan, where the bulk of Indus waters are consumed, as an example. As mentioned above, Pakistan had an average annual population growth rate of 2.61 % from 1961 to 2011. Moreover, an average rate of only 1.81 % from 2001 to 2011 indicates a slowing population growth rate. More importantly, total crop production grew by an annual average growth rate of 3.09 % for roughly the same time period, 1962–2010 (World Bank Statistics Database). Lest this be mistaken as growth in the production of inedible cash crops, the food production index during the same time period grew at an ever larger average annual rate of 3.43 %. Per capita caloric intake has increased from 1812 cal per person in the period 1961–1963 to 2340 cal per capita in the years 2001–2003 (Khan 2006, p. 16).

In addition to examining the food production data, it is also instructive to quantify the minimum water flow needed to sustain the current population of the Indus Basin. The United Nations declared a Human Right to Water on 28 July 2010, with the issuance of Resolution 64/292. A guideline provided in this resolution was that between 50 and 100 gal are needed per person per day to ensure the most basic needs, such as water for drinking, bathing, and other domestic uses. If we assume that roughly 300 million people live in the transnational Indus Basin (AQUASTAT 2011) and that each person requires 100 gal/day, this amounts to a minimum demand for about 92,000 acre-feet per year to sustain the entire current population of the

Indus or roughly 33.6 MAF/year. This is, of course, much less than the total estimated Indus surface water annual flow of about 200 MAF/year.

This data on food production in Pakistan and the total water requirements of the human population of the Indus suggests that there is more than enough water to take care of basic drinking and eating requirements. For this reason, it is simply not convincing that the root problem is "too many people" or "too little water." However, the per capita statistics I have presented are arrived at by mere arithmetic – they do not actually tell us anything about what is actually happening on the ground. As of 2008, an estimated 51 % of the population in Pakistan is experiencing food insecurity – which means consuming less than 2100 cal per day (Bhutta 2011, p. 24). That is to say, although there is more than enough water to go around for the current human population of "Industan," a large chunk of that population is experiencing acute material deprivation. Clearly, something is amiss. If absolute scarcity of food or water in the face of a growing population is not the issue, then what is?

The issue, as political ecologists and other critics of scarcity discourses have long argued, lies in the politics of distribution, allocation, and access, not aggregate supply and demand (Robbins 2011; Mehta 2010; Harvey 1974). There is often more than enough natural bounty available to meet the basic needs of the entire population – but the social and material infrastructure that distributes food and water across society is exclusionary. One could go so far as to argue that the "global water problem is neither one of physical water scarcity nor of excessive demographic development" and that it "is primarily the result of the fusion of the dynamics of water with the power of money in highly uneven ways" (Swyngedouw 2013, p. 828). The question is not how to increase the supply (or decrease the demand) of water; the question is how to share it in a more rational and egalitarian manner. The data on social classes in Pakistan reveals much about the distribution of resources in Pakistani society. For the task at hand, we can reduce the complex idea of "class" to differences in access to income. In 2008, the richest 10 % of the country's population held 56 % of the national income share, while the poorest 20 % held less than 10 % of the national income share. This is virtually unchanged from the situation 20 years earlier (World Bank Statistics Database). To speak of deprivation stemming from absolute water scarcity in Pakistan and the Indus without taking into account the extreme social (and geographical) differentiation within populations overlooks how people on the ground actually access the goods they need and want.

Amartya Sen, winner of the 1998 Nobel Memorial Prize for Economic Sciences, has articulated an influential alternative to studies of absolute scarcity that has come to be called the "entitlements framework." Sen's argument, in its original formulation, is that "famines can arise from causes other than food availability decline," (Sen 1977, p. 34). Sen explains that in a market economy, most if not all of the goods, including goods necessary for survival, are attained via market exchange. In this situation, it is crucial to look at the capacity of people to engage in useful exchange or otherwise secure the means of their social reproduction. This capability is not always related to the absolute supply of a desired good; it has as much to do with factors such as shifting terms of exchange (if what you produce suddenly loses value and thereby reduces your ability to exchange it for other goods), proximity to essential infrastructure, interventions of the state, and discriminatory exclusion based on factors such as age, gender, nationality, caste, class, or race. Sen's approach productively takes the focus away from physical scarcity and introduces an alternate vocabulary to discuss deprivation: social choice, entitlement, capabilities, and freedom.

These concepts from political ecology urge us to take social power dynamics seriously in considerations of environmental change. In deeply unequal and unevenly developed countries like India and Pakistan, access to land (which is mediated by factors such as caste and class) is arguably a much more relevant factor for evaluating water and food security than the total amount of water available (Mustafa et al. 2013). Access to water in the Indus Basin in Pakistan cannot be separated from access to land. Unlike in other arid regions, like the Western United States, where rights to water can extend beyond those who own riparian land, water rights in Pakistan has had everything to do with landownership since the region was brought under perennial irrigation by the British in the late nineteenth century. Indeed, all questions of structural resource scarcity and unequal social power in the Indus, and not just in the water sector, must at some point confront the question of agrarian land. As Marcus Taylor argues, "across the Indus watershed, ownership of land stands at the nexus of a series of power relations involving control over water, credit, and the labor of others" and that these "social relations are central to the international constitution of vulnerability as a socio-ecological relationship and greatly shape the ability of different social groups to adapt to hydroclimatic variations" (Taylor 2014, p. 135).

2.3 Structural Scarcity: British Canal Colonization and the Green Revolution

A historically sensitive characterization of land use patterns is a crucial component of understanding structural water scarcity because these patterns form the fundamental agro-ecological building block of Indus society. Arguably the first major structuring moment in this history occurred with the British project to irrigate the Indus plains (Akhter and Ormerod 2015; Gilmartin 2015; Ali 1988). This was an era of massive public investment in the Indus province of Punjab (today divided between Pakistan and India) under British colonial rule. From 1867 to 1892, cultivated area in Punjab increased 50 %, total kilometers of canals increased eightfold, railway mileage increased fourfold, and kilometers of roads doubled. From 1885 to 1947, the British built thousands of miles of canals in this arid landscape and facilitated the migration of farmers carefully selected by caste and kinship group. Captain C.H. Buck, a geographer with the Imperial Administration working in the Punjab, wrote in 1906 that "…I think I may truly state that the eyes of the masses have been opened to the great benefits they are receiving under British rule" (Buck 1906,

p. 67). The resulting "canal colonies" of southwest and central Punjab were the centerpiece of a massive socioecological colonial engineering project.

The ecological project to make the Indus Basin an irrigated breadbasket was also necessarily a social project to create a stable- and loyal-classed society at the heart of British India. As Imran Ali (1988) documents, the British development of the canal colonies favored certain segments of society with land grants. Favored cultivators, who eventually managed to secure formal property rights over land (and thus water), tended to belong to traditionally dominant agricultural castes, such as Jats, Arain, and Rajputs. Thus large-scale dispossession of the indigenous people of the dry inter-riverine tracts of Punjab, pejoratively called *jangalis* (jungle dwellers), accompanied the establishment of a new agrarian order centered around canals (Gilmartin 2015; Gazdar 2011; Ali 1988; Nazir 1981). The actual operation of the canal system, as it was conducted by local administrators and engineers, also played out in favor of landed notables in the area (Gilmartin 2015; Ali 1988), and the system continues to operate in this way today (Mustafa 2013; Javid 2011).

Since the Green Revolution, a technological transformation of agriculture that swept across North India and Pakistan in the 1960s, the rates of landlessness and farm consolidation have been skyrocketing. The most emphasized aspect of the Green Revolution is that it increased the production of food: the output of wheat in Pakistan increased by 91 % and rice by 141 % between 1960 and 1970. The Green Revolution in Pakistan came in two phases: 1959–1964 and 1965–1970. In the first phase, agriculture grew at a rate of 3.7 %, driven mainly by an expansion in irrigation facilities, most notably tube wells. Agricultural production grew at an even faster pace in the second phase, at a rate of 6.3 %, propelled by the technologies of high-yielding variety seeds, chemical fertilizers, and pesticides. Although a given unit of land was producing more food, it would nevertheless be mistaken to assume increased agricultural productivity. For that to be true, output would have to increase with respect to all inputs and the Green Revolution depended critically on increased inputs.

But these advances in production depended on the increased availability of water. In the first 5 years of the 1960s, about 25,000 tube wells were installed to double the farm area serviced by tube wells. Even the new seed technology was dependent on the timely application of water to have effect. Between 1964/1965 and 1969/1970, 45,000 more tube wells were installed, increasing the area serviced by tube wells sixfold over the course of a decade. Overall, the enhanced availability of water, either through private tube wells or public canals, is estimated to have contributed about half of the total increase in agricultural output during the Green Revolution in Pakistan. Fertilizer inputs also increased dramatically; 235 % between 1965/1966 and 1971/1972 (Zaidi 2006, pp. 28–29).

The distributional impact of the Green Revolution was highly uneven across classes. One manifestation of this was the decline in tenurial security for cultivators. As agricultural production became more lucrative, landowners decided to dismiss their tenants, sink money into technological investments, and engage in what is officially referred to as "self-cultivation." In reality, this involves hiring wage workers for the work of cultivation, but without tenurial rights. The number of tenant farms was cut by half between 1960 and 1990, and the landless rural population in Pakistan swelled to between 30 and 49 million or about 40 % of the rural population. Indeed, between 1960 and 1990, concentration in landownership has increased and rural household income and secure employment for a large segment of the population has fallen (Niazi 2004). Today, nearly half of the total area of the basin is owned by about 2 % of households (Condon et al. 2014, p. 65).

2.4 Water Supply and Security in the Context of Climate Change

Thus far I have examined both absolute and structural scarcity in the Indus Basin. I've argued that while declines in absolute scarcity are alarming, a nuanced political ecological perspective must focus on the interaction between absolute and structural scarcity. This political ecological understanding of the water scarcity-security nexus helps us think about the future of water security in the Indus region more critically. Recent tree-ring reconstruction research in the Upper Indus Basin area (around Swat, Pakistan) presents a deeper temporal perspective on the annual variability of flows in mainstem Indus (as distinct from the entire Indus River System, which consists of the Indus and all its tributaries). Tree-ring reconstructions are useful for the study of inter-annual variability of river flows. This is because they permit the construction of an average figure for lows over a much longer temporal scale, allowing one to evaluate with greater certainty whether an observed deviation is statistically aberrant or not. Cook et al. (2013) examine tree-ring data from several tree species near the Partab Bridge gauging station, as shown in Fig. 2.2, selecting the actual recorded flows from 1975 to 2004 as a calibration period. The stream flows they reconstruct are for the 556 year period between 1452 and 2008. What is relevant is not necessarily the level of discharge this reconstruction shows us – but the high degree of annual variation around the mean flow.

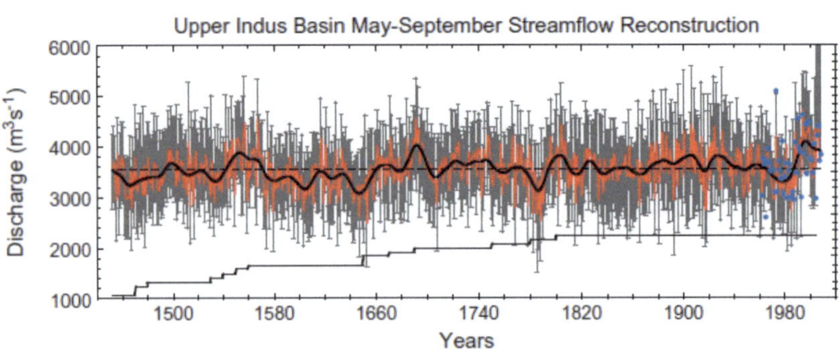

Fig. 2.2 Upper Indus Basin stream flow reconstruction from Cook et al. (2013), p. 7

Cook et al. make the critical observation that there are long periods of sustained low flow on the Indus mainstem. For example, flow during the 111 year period (1572–1683) was on average 8 % lower than the 556 year average. And during a smaller 26 year period (1637–1663) the flow was even lower – on average 11 % lower than the 556 year average. The authors warn "should either of these low-flow periods repeat in the future, the resulting cumulative deficit could seriously reduce Pakistan's capacity for irrigation and hydroelectric power generation provided by the Tarbela Reservoir and Dam" (Cook et al. 2013, p. 9). It is important to note that both of these low-flow periods occurred before human industrial activity could have had anthropogenic impacts on climatic change – so the effects of greenhouse gas emission over the past decades creates uncertainty that cannot be wholly allayed by reconstructions of climate history. Two areas of water supply in the Indus are especially prone to uncertainty – the timing and magnitude of annual precipitation and the rate and significance of deglaciation in the Himalayas and sub-Himalayan mountain ranges.

The Indus Basin receives an average of 405 mm (17.9 in.) of mean annual precipitation, which, for the most part, ultimately either percolates to the aquifer or runs to the sea as surface runoff. Rainfall in the Indus Basin is highly concentrated temporally, falling largely in monsoon summer months. There is less controversy over how climate affects monsoonal patterns than over glacial melt, but this is likely because there is not enough data to run Global Circulation Models to account for regional variation, nor are monsoonal dynamics understood well enough at this point to model them. Christensen et al. (2007), writing for the Working Group of the Intergovernmental Panel on Climate Change, forecast an increase in the number of rainy days in South Asia, as well as an increase in the number of extreme events as a result of climate change.

More precipitation seems like an antidote to the threat of water scarcity, but it might not have the same implications for water security. More extreme precipitation events could lead to more severe hazards in Pakistan (Briscoe and Qamar 2006). Indeed, while much of the water in the Indus comes from glacial melt, the contributions of rainfall are more intensely concentrated into a shorter period of time.

The total amount of water available in the Indus system depends ultimately on levels of precipitation. But the timing of the availability of this precipitation depends on other factors – such as the amount of precipitation that is captured by glaciers in the winter months and how much is released as meltwater in the summer months. Although glaciers cover only 20,324 km^2 or 1.78 % of the total area of the Indus River system basin (Kaser et al. 2011, p. 20,226), they serve an important function in storing water during the winter months and releasing it in the summer. This dependence on the regulative aspects of glaciers is shared by the Indus Basin and other arid regions (Condon et al. 2014; Kaser et al. 2011). While scientists agree on the crucial contribution of glacier meltwater in the Indus, the question of what we can expect from this source of water in the context of global climate change is a subject of debate. Some analysts argue that retreat of glaciers in the Western Himalayas and the Karakoram Mountains in the north of Pakistan has already begun, and the eventual outcome will be reduced river runoff preceded by a period of above-average flows (Condon et al. 2014; Briscoe and Qamar 2006, p. 27).

To sum up, the only certainty regarding future water supplies in the Indus is that they will be marked by great uncertainty. We do not know yet how anthropogenic climate change will impact the onset and magnitude of climatic events like the South Asian monsoon or rates of deglaciation. We do know that the Indus mainstem (and, by implication, the entire Indus River System) has gone through long periods of greatly reduced surface water flows. To consider how this impacts water security in India and Pakistan, it is important to understand how the confluence of history, political economy, and technology actually determine the uneven allocation and distribution of water in the Indus Basin.

Returning to the above discussion of the British canal colonization of the Indus valley and the impacts of the Green Revolution on the region, we must keep in mind that Pakistan and northwest India – the populous heartland of Industan – are perhaps the prime examples in the world of what has been called "hydraulic societies" (Gilmartin 2015; Briscoe and Qamar 2006; Ali 1988). This is because of these regions' overwhelming economic and social dependence on irrigated agriculture. In the case of Pakistan agriculture accounts for a quarter of the country's GDP and directly or indirectly provides 60 % of the population with their livelihood. The agrarian economy and culture of Indian Punjab likewise exerts an outsized influence on the food supply and political economy of India. Looking simply at the provinces of Indian and Pakistani Punjab is revealing in this respect. In 2005–2006, Pakistani Punjab accounted for 77 % of national wheat production in Pakistan (Bureau of Statistics 2008, p. 93), and the corresponding number for Indian Punjab with respect to India for the years 2006–2007 is 57 % (Government of Punjab 2008, p. 169).

All this is to say that water security in the transnational Indus must be first and foremost concerned with agrarian political economy. Others have rightfully called for more attention to the ways in which legal and administrative institutions for the allocation of water in the Indus Basin should be revisited and reformulated (Condon et al. 2014; Yang et al. 2014). But it must be kept in mind that Industan is first and foremost, as far as water is concerned, a hydraulic society that is overwhelmingly structured by the necessities of irrigated agriculture in arid lands. So too are the social structures in these regions that reproduce unequal access to water shaped decisively by agrarian political economy. Thus the historical and political economic factors such as canal colonization and the Green Revolution – which affected the transnational Indus and especially the transnational region of Punjab – must be understood as the medium through which absolute water scarcity will be experienced and felt. Policy options for tackling water security in the future must therefore begin with the structures of agrarian political economy – in other words, with reforming the current structure of property in land. This is in contrast to the main policy formulations which tend to see expertise and technology as a way to achieve water security in the Indus (for an alternative framing of the politics of water infrastructure on the Indus, see: Akhter 2015a, b, c; Akhter and Ormerod 2015). Indeed, there have been a growing number of calls to pay closer attention to the connection between land ownership, water security, and food security in the Indus region (Taylor 2014; Gazdar 2011; Toor 2010).

Land reform was legislated in Pakistan in 1959, 1972, and 1977 and in India in the early 1950s and 1972. Although implementation in both countries was less than ideal, Indian Punjab today has less inequality of distribution of land among farmers (Zaidi 2006; Sims 1988). Thus the case for land reform as a way to tackle the broader structures of inequality that shape water security in the Indus is even stronger in the Pakistani portion of the Indus than it is in India. Saadia Toor's (2010, pp. 109–110) conclusions after her analysis of the connections between land ownership and food insecurity in the Pakistani context are worth quoting at length:

> A radical program of land redistribution is required in Pakistan to address issues of food (in) security in particular and of poverty in general. Experts argue that land redistribution along the lines of greater equity will also take care of many of the problems that the agricultural sector is facing, including a fall in productivity, waterlogging and salinity, and wastage of water. The current highly skewed system of land ownership is both the cause and the effect of power. The existing system gives large landowners the ability to exploit smaller landowners and deny them their due share of resources, especially water. Not only are large landholdings less productive than smaller ones, but large landowners in Pakistan also engage in rent-seeking rather than productive activity.

2.5 Conclusion

The historical and political economic context through which water demand and supply are established gives us the means to reframe the water scarcity debate. Water scarcity is often presented as an imbalance between demand and supply. While this is technically correct, I have tried to show that a deeper understanding of water scarcity must historically contextualize systems of demand and supply. In other words, water supply and demand are not exogenous variables; rather, the magnitude and nature of supply and demand in any given region are produced as a part of a region's history and geography. And, as we have seen, the very categories of analysis that we are used to working with are no longer sufficient when we study the historical geography of the Indus in detail. I have also drawn attention to the importance of historical forces that act transnationally. For example, the Green Revolution trends of land and water use highlighted above were not limited to the Pakistani side of the Indus Basin. These very same technologies and agricultural management strategies also transformed agricultural production methods and political economy in Indian Punjab in the eastern portion of the Indus Basin. Therefore, when selecting categories with which to study water security, we should be cautious of analyses that attribute causal efficacy to physical scarcity and instead be attuned to the complex and place-specific links between absolute scarcity and the national and transnational conditions that create and reproduce structural scarcity.

References

Akhter M (2015a) The hydropolitical Cold War: the Indus Waters Treaty and state formation in Pakistan. Polit Geogr 46:65–75

Akhter M (2015b) Infrastructure nation: state space, hegemony, and hydraulic regionalism in Pakistan. Antipode 47(4):849–870

Akhter M (2015c) Dams as a climate change adaptation strategy: geopolitical implications for Pakistan. Strateg Anal 39(6):744–748

Akhter M, Ormerod KJ (2015) The irrigation technozone: state power, expertise, and agrarian development in the US West and British Punjab, 1880–1920. Geoforum 60:123–132

Ali I (1988) The Punjab under imperialism, 1885–1947. Princeton University Press, Princeton

AQUASTAT (UN Food and Agriculture Organization Information System on Water and Agriculture) (2011) Indus basin water report 37, at: http://www.fao.org/nr/water/aquastat/basins/indus/index.stm. Accessed on 24 Sept 2015

Bhutta Z (principal investigator) (2011) National nutrition survey Pakistan

Briscoe J, Qamar U (2006) Pakistan's water economy: running dry. The World Bank, Washington, DC

Buck CH (1906) Canal irrigation in the Punjab. Geogr J 27(1):60–67

Bureau of Statistics, Government of Punjab (2008) Punjab development statistics, 2007. Punjab Government Printing Press, Lahore

Central Ground Water Board (2014) Ground water year book, 2013–2014. Ministry of Water Resources, Government of India. Available here: http://www.cgwb.gov.in/documents/Ground%20Water%20Year%20Book%202013-14.pdf. Accessed 30 Oct 2015

Christensen JH, Hewitson B, Busuioc A, Chen A, Gao X, Held I, Jones R, Kolli RK, Kwon W-T, Laprise R, Magaña Rueda V, Mearns L, Menéndez CG, Räisänen J, Rinke A, Sarr A, Whetton P (2007) Regional climate projections. In: Solomon S et al (eds) Climate change 2007: the physical science basis. Contribution of Working Group I to the Fourth Assessment Report of the Intergovernmental Panel on Climate Change. Cambridge University Press, Cambridge

Condon M, Kriens D, Lohani A, Sattar E (2014) Challenge and response in the Indus Basin. Water Policy 16(S1):58–86

Cook ER, Palmer JG, Ahmed M, Woodhouse C, Zafar M, Wahab M, Khan N (2013) Five centuries of upper Indus River flow from tree rings. J Hydrol 486:365–375

Gazdar H (2011) The fourth round and whey they fight on: an essay on the history of land and reform in Pakistan. In: Leveling the playing field: a survey of Pakistan's land reforms. PANOS South Asia, Kathmandu, pp 8–65

Gilmartin D (2015) Blood and water: the Indus River basin in modern history. University of California Press, Oakland

Government of Pakistan (2005) State of the environment report 2005 (Draft). Available at: http://environment.gov.pk/state-of-environment-report/. Accessed 30 Aug 2015

Government of Punjab (2008) Statistical abstract of the Government of Punjab. Retrieved from http://www.pbplanning.gov.in/. Accessed 26 May 2016

Harvey D (1974) Population, resources, and the ideology of science. Econ Geogr 50(3):256–277

Javid H (2011) Class, power, and patronage: landowners and politics in Punjab. Hist Anthropol 22(3):337–369

Kaser G, GroBhauser M, Marzeion B (2011) Contribution potential of glaciers to water availability in different climate regimes. Proc Natl Acad Sci 107(47):20223–20227

Khan MH (2006) Agriculture in Pakistan: change and progress 1947–2005. Vanguard Books, Lahore

Laghari A, Vanham D, Rauch W (2012) The Indus basin in the framework of current and future water resources management. J Hydrol Earth Syst Sci 16:1063–1083

Malik B (2011) Save water save Pakistan. Ferozsons, Lahore

Mehta L (ed) (2010) The limits to scarcity: contesting the politics of allocation. Routledge, New York

Mustafa D (2013). Water resource management in a vulnerable world: the hydro-hazardscapes of climate change. Philip Wilson Publishers, London
Mustafa D, Akhter M, Nasrallah N (2013) Understanding Pakistan's water/security nexus. United States Institute of Peace, Washington, D.C.
Nazir P (1981) Transformation of property relations in the Punjab. Econ Pol Wkly 16(8):281–285
Niazi T (2004) Rural poverty and the Green Revolution: the lessons from Pakistan. J Peasant Stud 31(2):242–260
Robbins P (2011) Political ecology: a critical introduction. Wiley, Malden
Sen A (1977) Starvation and exchange entitlements: a general approach and its application to the great Bengal famine. Camb J Econ 1:33–59
Sims H (1988) Political regimes, public policy, and economic development: agricultural performance and rural change in the two Punjabs. Sage Publications, New Delhi
Swyngedouw E (2013) UN water report 2012: depoliticizing water. Dev Chang 44(3):823–835
Taylor M (2014) The political ecology of climate change adaptation: livelihoods, agrarian change and the conflicts of development. Routledge, New York
Toor S (2010) The structural dimensions of food insecurity in Pakistan. In: Kugelman M, Hathaway R (eds) Hunger pains: Pakistan's food insecurity. Woodrow Wilson International Center for Scholars, Washington, DC
Yang YCE, Brown C, Yu W, Wescoat J, Ringler C (2014) Water governance and adaptation to climate change in the Indus River Basin. J Hydrol 519:2527–2537
Zaidi SA (2006) Issues in Pakistan's economy, 2nd edn. Oxford University Press, New York

Part II
The Costs and Scale of Transboundary Conflict

Chapter 3
Water Insecurity in the Indus Basin: The Costs of Noncooperation

Ashok Swain

Abstract This chapter examines the international and domestic costs of noncooperation among and within the countries comprising the Indus basin. Its focus is on all four of the co-riparian basin states, but India and Pakistan receive special attention due to their prominent position in the basin's water use and management. Besides the international costs, including intensification of traditional enmities, heightened distrust, persistence of territorial disputes, terrorism, and weakening of regional norms of cooperation, this chapter also examines the way in which noncooperation in regard to water resources plays an instrumental role in creating and aggravating the interprovincial and interethnic divisions within basin countries. The bilateral and internal mistrust and suspicions in the basin have restricted the ability of riparian countries to develop critical water development projects like Tulbul Navigation project/Wullar Barrage, Sutlej Yamuna Link Canal, and Kalabagh Dam to make best possible use of available water resources, and this in turn has impacted negatively on the social and economic development of these countries.

Keywords Indus Waters Treaty (IWT) • Tulbul Navigation project • Wullar Barrage • Sutlej Yamuna Link Canal • Kalabagh Dam

3.1 Increasing Demand for Water in the Indus Basin

Water plays a fundamental role in peace and development of a society. Disparity in access to the water resource as well as its misuse and mismanagement destabilizes economy and welfare, affects human security, and creates risk of violent conflict. Thus, water scarcity is not limited to issues concerning environment and development only; it has become part of the political agenda of many nations and an important national and regional security issue. The South Asia region is highly populated and also houses a large number of the world's poor. In recent years, besides the rapidly increasing population, a large part of the subcontinent is also experiencing

A. Swain (✉)
Department of Peace & Conflict Research, Uppsala University, Uppsala, Sweden
e-mail: ashok.swain@pcr.uu.se

Z. Adeel, R.G. Wirsing (eds.), *Imagining Industan*, Water Security in a New World, DOI 10.1007/978-3-319-32845-4_3

rapid but likely unsustainable economic growth. The water scarcity situation has already become severe in the region, where people are highly dependent on this natural resource for their livelihoods.

The South Asia region does not have a carefully crafted integrated water resource development strategy to address the growing water scarcity situation. Most of this region's large river systems flow across national boundaries and have already become sources of interstate conflicts. Global climate change impacts have also brought serious risks to the accessing of water resources in South Asia (Earle et al. 2015). In particular, the impact of climate change on glaciers directly affects water flow in many of the major international rivers in this region, especially the ones originating from the Himalayas. In the Himalayan river systems, due to the increased rate of glacial melt, the water flow in spring has increased and water flow during the remaining seasons – especially during summer when it is most needed – has decreased.

The Indus River basin is one of the largest river basins in South Asia. It covers an area of 1,165,000 km^2 (Swain 2004) spreading across four countries – China in the northeast, India in the east, Afghanistan in the northwest, and Pakistan, including Khyber Pakhtunkhwa and a majority of the plains of Punjab and Sindh (Rehman and Ahmed 2005). Immediately after independence, both India and Pakistan came into conflict over their share of water in the Indus system. After more than 8 long years of hard negotiation, both the riparian countries reached an agreement on 19 September 1960. The Indus Waters Treaty (IWT) was signed at Karachi by India's Prime Minister Jawaharlal Nehru and Pakistan's President Mohammad Ayub Khan. The 1960 Treaty virtually was an extension of the 1947 partition process as it allocated the three eastern rivers of the basin (the Ravi, Beas, and Sutlej) to India and the three western rivers (the Indus, Jhelum, and Chenab) to Pakistan. The Treaty allowed Pakistan to construct a system of irrigation works on the western rivers to compensate for the loss in irrigation supply from the eastern rivers. The Treaty also paved the way for building two major dams in Pakistan: the Mangla Dam on the Jhelum River and the Tarbela Dam on the Indus. Pakistan was also able to construct several projects to divert water from its western rivers to replace reduced flows in the Sutlej Valley Project Region. On the other hand, the Treaty permitted India some storage capacity for the reservoirs to be constructed on the western rivers. Thanks to the Treaty, India constructed the Bhakra Nangal and Beas projects and the Indira Gandhi Canal project enabling the use of all of the allocated water from the eastern rivers.

The Indus Waters Treaty has stood the test of time for the last 55 years, though during this period many questions have been raised and both states have expressed differences on many issues. The World Bank showcases this Treaty as a major success story, proof of its credibility as a third party negotiator in any resource dispute. Water peace researchers highlight this water treaty to show that water has the potential to bring even two adversaries like India and Pakistan to a platform of cooperation. It is true that the Treaty has survived two major wars between India and Pakistan (1965 and 1971) and an undeclared war in 1999. Though the Treaty is over river waters, it is designed primarily to split the river system into two, not to provide for the sharing of its waters. David Lilienthal had envisaged the agreement to be one that would treat the whole basin as a single unit so that the two riparian states could cooperate in more effective ways for the sustainable management of river and land resources. But the Indus Waters Treaty, in facilitating the partition of the river system between the two riparian countries, has in fact contributed to reducing the scope of engagement between them. The Indus Waters Treaty is not a marriage of two consenting adults to lead a life together but a mutually agreed divorce settlement. Though the best possible use of Indus waters needs both the major riparian countries to work together for the construction of water infrastructure and pollution control, the IWT unfortunately does not encourage any development toward joint basin management (Swain 2004).

Pakistan is one of the world's most arid countries, having an average rainfall of less than 240 mm (Briscoe and Qamar 2006). Water from the Indus River system is the primary source of water for the country's domestic agriculture and industrial needs. Pakistan's agriculture is heavily dependent on surface water irrigation because of low precipitation and limited availability of quality ground water. More than three-fourths of Pakistan's agricultural land area is supported by irrigation, and the country ranks second in the world for its irrigation dependence after Egypt (Ahmed et al. 2007). Of the total irrigated area supported by the Indus system, 74 % is located in Pakistan (Laghari et al. 2011). There is no doubt that the Indus River system is a major source of water for Pakistan; but its water is also extremely critical for the water-scarce but agriculturally developed northwestern part of India (Swain 2009). In the arid northwestern region of India, which has become the country's breadbasket in the post-Green Revolution period, the Indus River provides the economic foundation (Zawahri 2009). With growing populations in both India and Pakistan, there is increasingly higher demand for food and energy, which adds to the pressure on the limited water resources of the Indus River system.

3.2 Mistrust Between India and Pakistan and the Tulbul Navigation Project/Wullar Barrage

Under the 1960 Treaty, India and Pakistan have established the Permanent Indus Commission to supervise the sharing of Indus water. The commissioners meet at least once a year, alternating their meeting places between the two countries. During these meetings, they exchange data and information regarding the ongoing and planned water development projects in the basin. If either of the countries has questions or objections regarding any construction or maintenance project, the issue can be referred to the Commission, and if the Commission fails to resolve the issue, then the matter can be referred to a neutral expert, appointed either by the two members of the Commission or by the World Bank (Uprety and Salman 2011). If the verdict of the neutral expert fails to satisfy the parties, then the issue gets treated as a dispute, in which case an International Court of Arbitration could be established to address it (Swain 2009).

The geographical setting of the Indus River system has facilitated the partitioning of the six rivers in equal numbers between the two riparian countries under the 1960 Treaty. It is not that India harbors intentions of this sort, but the mountain terrain and intractable conflict over Kashmir have thwarted any Indian plan to divert the waters of the western rivers southward to its mainland. However, any Indian plan to develop the water resources of the portions of the three western rivers lying within the Indian-controlled sector of Kashmir has always created serious concern for Pakistan. India consistently maintains that it has not built any storage facilities on western rivers. However, there have been a number of acrimonious disputes between India and Pakistan over Indian projects on these rivers.

Pakistan objected, for instance, to India's construction of the Baglihar hydroelectric dam on the Chenab River. The dispute was finally settled in 2007 by the verdict of a World Bank-appointed neutral expert. Another major dispute arose over India's Kishanganga hydroelectric project on the Kishanganga–Jhelum River. Pakistan took the case to the International Court of Arbitration. The Court, in its final order on 20 December 2013, ruled in favor of India's plan to build the project, but at the same time, India was asked to release nine cusecs of water to maintain a minimum flow in the river, and also, India should not in the present project and all future projects on western rivers use the "drawdown flushing" technique to deal with sedimentation (Desai and Sindhu 2014). Among other Indian planned projects being objected to by Pakistan are the Sawalkot hydroelectric project and the Dul Hasti hydroelectric project on the Chenab River. However, the oldest major dispute over the interpretation of the Treaty in regard to Indian use of the western rivers, dating back to the 1980s, arose over the Tulbul Navigation project (named by Pakistanis the Wullar Barrage). This dispute has not been settled yet.

India plans to build a 439-ft long, 40-ft wide barrage on the River Jhelum at the mouth of Wullar Lake, a natural lake in Kashmir Valley. According to this plan, the barrage is expected to have a maximum storage capacity of 0.30 million acre-feet of water. With the help of this Tulbul Navigation project, India envisions regulated release of water from the lake during the lean season months of October to February to maintain a minimum draught of 4.5 ft in the river to facilitate year-round navigation between Srinagar and Baramulla for commercial trade and tourism. The Jhelum River provides an important means of transport for goods and people in the Kashmir Valley, and to facilitate navigation throughout the year, a minimum depth of water is needed. The barrage project was conceived in the early 1980s and work began in 1984.

Pakistan, after learning of the project from media reports in 1984, opposed it on grounds that upstream water storage would affect the water supply to its own Upper Chenab Canal, Upper Jhelum Canal and Lower Bari Doab Canal. India contends that development of the Tulbul Navigation project is permissible under the Indus Waters Treaty of 1960, which allows for the nonconsumptive use of navigation, while Pakistan maintains that the project is a "storage project" and finds it in violation of the Treaty (Mirza 2013). However, India maintains that the barrage would not reduce the quantum of water flow to Pakistan. Being a "control project," India argues that the project would even be beneficial to Pakistan for regulating water flow to the Mangla Dam and would also support flood control and improve hydropower generation.

However, Pakistan argues that with the help of the barrage, India would control the flow of water into the Jhelum, creating drought and flood situations at will downstream and, thus, be in position to threaten Pakistan's agriculture. The real worry of Pakistan is that this project can bring strategic advantage to India by controlling the river's upstream flow. Pakistan has an apprehension that India might use the barrage to dry the riverbed during periods of war to facilitate troop movement to Pakistan. This apprehension is based on experience gained in the 1965 war, when the Indian army at one point had failed to advance because of high water depth in the Bambawali-Ravi-Bedian Link Canal. Pakistan brought the barrage issue to the Indus Waters Commission in 1986, but the Commission was unable to resolve it. Pakistan then wanted to take the case to an International Court of Arbitration (ICA), a proposal which India rejected by responding officially with postponement of construction. India stopped working on the project in 1987, but it has since then frequently been pressed to restart construction (Misra 2010).

After many rounds of talks, the Tulbul issue remains yet unresolved. India has agreed to modify the barrage structure's design and also to maintain a fixed crest level of river water flow (Zawahri 2008). The latest round of talks over this issue took place between the delegations of the two countries in New Delhi on 27--28 March 2012. In these talks, Pakistani officials have asked India to provide more information about the project. At the same time, there have been suspicions in Pakistan that India has surreptitiously been continuing to work on a structure associated with the Tulbul Navigation project under cover of the Jhelum Flood Control and Prevention Scheme. On 27 August 2012, a militant attack on the laborers working for the project brought the matter to the public domain (Wani 2012). India argues that the suspension of barrage work is harming the interests of the people of Kashmir and also depriving the people of Pakistan of irrigation and hydropower and flood control benefits. Besides navigation benefits, the barrage project will certainly contribute to the India's energy sector because a better discharge during lean season will add to the underutilized capacities of the downstream hydropower projects at Uri. A regulated flow during lean season would also help Pakistan in getting more hydropower from its Mangla Dam. However, Pakistan's lack of trust about India's real intentions still stands as the main hurdle in the way of implementing this project.

Both India and Pakistan realize the importance of the Indus waters for their own food security. Thus both countries have a strong national interest in maintaining institutional sharing of these waters. However, the deep mistrust between India and Pakistan has thwarted the implementation of a number of seemingly advantageous water development projects in the basin, of which the Tulbul/Wullar project is only one of the more conspicuous. It has all the ingredients to provide a win-win situation for both the riparian countries. Still it has been stalled for nearly three decades.

3.3 Internal Politics and Unfinished Water Projects in India and Pakistan

Nineteenth-century colonial India witnessed a major growth in the Indus basin's system of irrigation, flood control embankments, and drainage networks (D'Souza 2006). This colonial hydrology changed many existing relations between land and water not just in the sphere of traditional and colonial water technology but also in established economic and political and territorial relations. These have undergone massive transformations since decolonization. The Indus system is not just the most important but also the only source of irrigation for Pakistan, which is still primarily an agricultural country. India's state of Punjab, just across the border, is called the food bowl of India. Thus, the Indus system supports a large portion of the water-intensive agricultural lands in both countries. Moreover, nearly 30 % of the world's cotton supply comes from the Indus basin region shared by India and Pakistan; and the cotton cultivation takes an average of 737 billion gallons of water annually from the Indus (equivalent to all of Delhi's residents household water needs for more than 2 years) (Wheeler 2011).

After the Partition of India in 1947, the territories of Pakistan became the lower riparian and India as the upper riparian. The Indus Waters Treaty of 1960 gave exclusive rights of the three western rivers and the three eastern rivers to India. Pakistan received over 80 % of the share while India receives the rest. However India being the upper riparian controls most of the water flows to Pakistan. Some growth models predict that by 2025, India and Pakistan's populations will be more than tripled as to what it was when the Treaty was signed (Wheeler 2011). The Treaty does not have any roadmap for "future water cooperation" and it is also silent on the issue of water quality. In the recent years, the increasing scarcity of water and its increased use shifting not just from agriculture to domestic and navigation purposes but to secondary purposes such as a coolant for nuclear reactors and thermal power plants have led to diminishing flows for both countries. There are several bilateral concerns, but the primary source of disputes has been over the building of dams. Pakistan is especially concerned of India's plan to build 155 hydel projects on the Indus basin that could significantly diminish flows downstream toward Pakistan (Bhutta 2011).

Pakistan claims that several of the larger dams that India has built or plans to build violate the Indus Waters Treaty, while India claims that the dams built on their side of the river do not violate the provisions of the Treaty and that Pakistan has poor water management facilities in place despite withdrawing more than the fair share of the water from the system (Sumbal 2014). However these water issues flow into non-water issues between India and Pakistan regarding manipulation of water by India, using it as political leverage over the Kashmir issue or ethnic rifts between Punjab and Haryana. The Khalistan movement in Punjab, India, and the demand for independence in Indian-administered Kashmir are flashpoints in this interconnectivity between international relations, ecological imbalance, social inequality, and state security.

Water sharing disputes have a history of provoking internal conflicts between different ethnic groups in India (Swain 1998). After the signing of the Indus Waters Treaty, Ravi, Beas, and Sutlej rivers were used exclusively by the Indian state of Punjab. However with the reorganization of states in India in 1966 and the creation of the state of Haryana, internal disputes started over various water-related issues between Punjab and its neighboring states. Punjab that had once been the riparian of Yamuna, Beas, and Sutlej now needed to share the waters not just from the Indus system, but it lost its riparian status for Yamuna as well which now ran only through Haryana. In 1976 Haryana received water from Ravi River, but Punjab was denied water from Yamuna River. One of the most controversial water sharing disputes till today between these two states remains over the construction of the Sutlej Yamuna Link Canal. Once this 214 km canal gets completed, it would provide water to Haryana from the Sutlej River. When its construction began in 1984, Punjab was already at the peak of a violent movement in demanding independence of the state constituting Sikh ethnic majority called Khalistan. Protests under the leadership of Sant Harchand Singh Longowal were organized against the canal, and militants killed a number of laborers and engineers engaged in the construction of the canal (Dhaliwal 2014).

Designed to transport 10,500 cusecs of water, the canal is going to provide 6,500 cusecs of water to Haryana. Besides carrying water to a part of Haryana, it will also bring irrigation possibilities to some villages of Ropar and Patiala districts of Punjab. Two small hydropower plants have also been planned along the canal to generate power. The portion of the canal in Haryana's territory has been completed, but the work has remained incomplete on the Punjab side. In 2004, the Assembly of state of Punjab passed the Punjab Termination of Agreements Act 2004 to avoid the completion of the canal in its territory. The Termination of Agreement Act created serious concern at the national level forcing the central government to intervene on the issue. The incomplete canal project in Punjab now is in ruins, and, in the rainy season, the canal turns into a big water channel flooding agricultural fields.

In the case of Indian-administered Kashmir, water sharing disputes and the demand for independence are even more strongly intertwined. The international dimensions of the conflict originated in 1947 when Kashmir was divided, giving Pakistan the control of what they call Azad (free) Kashmir, while India remained in control of the remaining two-thirds of the state. This division led to constant territorial disputes and subsequently the formation of the Line of Control (LOC) in 1972. However, despite the LOC, neither the territorial nor the religious conflicts have abated in this Muslim majority state in Hindu majority India. Add to this, the problem of shared river water resources takes on new dimensions internally for India. In pursuit of augmenting the power shortages in the nation, the central Indian government is building a series of small hydroelectric dams on the western rivers of the Indus system in Indian-administered Kashmir. However, much of the power generated is distributed to the national Indian grid, often leaving the conflict-ridden state without adequate power, a situation which is especially difficult during the harsh winters (Arnoldy 2010).

Kashmiris believe that Jammu and Kashmir state has the potential to produce 20,000 MW of hydropower, enough to make the state energy self-reliant and an independent Kashmir a viable option (Arnoldy 2010). Thus, the demand to reclaim for the state power projects run by an Indian central agency has become part of the demand for separatist forces. The Kashmiris' main concern is that the Indus Waters Treaty does not allow the harnessing of the full hydropower potential of the state. Extremely energy starved, the state government is trying to put pressure on the Indian government to compensate for the loss it is incurring due to IWT obligations. Kashmir's three main rivers have been allocated for Pakistan's use in the Indus Waters Treaty, thereby encouraging resentment. The state legislature passed a near unanimous resolution in 2002 calling for a review and annulment of the Indus Waters Treaty.

For a state already torn between the Hindu-Muslim nationalistic rhetoric between Pakistan and India, the IWT's appropriation of rivers represents the theft of local resources and yet another reason to demand independence, thus making water and electricity not just an issue of intrastate conflict between India and Kashmir but also an interstate one between India and Pakistan. Pakistan's fear that India could use its "water weapon" to put Pakistan in a drought-induced famine has its parallel in Pakistan's often being blamed for aiding insurgency and stoking separatist demands among the already agitated youth of Kashmir. Anti-India nationalists and militant groups based in Pakistan have started using the water issue as a fresh source of hatred. In particular, Hafiz Saeed, Jamaat-ud-Dawa leader, regularly uses the water dispute to criticize India in his Friday sermons (Polgreen and Tavernise 2010).

In Pakistan, Sindh and Punjab have been in dispute over their rightful shares at least from the mid-nineteenth century onward, from the time when colonial administration started the canal construction (Mustafa 2010). Balochistan is also in dispute with Sindh as a lower riparian of the Khirther Canal. The Pakistan Apportionment Accord of 1991 did not have any water allocation for any of the special (non-provincial) areas of the country, including Gilgit-Baltistan, Federally Administered Tribal Areas, Pakistan-controlled Kashmir territory, and Islamabad, the capital city of Pakistan. The increasing population and growing economy of these regions have brought water scarcity to these areas, prompting their leaders to start demanding a fair share of Indus water.

The four provincial governments, basing their calculations on the average flow of 114.35 million acre-feet (MAF) of the Indus system, agreed in 1991 on the allocation of 55.94 MAF to Punjab, 48.76 MAF to Sindh, and the rest (9.45 MAF) shared between Khyber-Pakhtunkhwa and Balochistan provinces. This accord suffered from a crisis of legitimacy as it was concluded at a time when a single political party was in power at the central level and also in all the four provinces of Pakistan for the first time. The water sharing issue is still simmering between Punjab and Sindh, the two powerful provinces in Pakistan. Punjab keeps staking its claim on the ground of "historical use," while Sindh complains of getting less water than its entitlement (Ahmad 2012). A quite common perception in Sindh is that the Sindh compensates Punjab for water that Pakistan negotiated away to India in the IWT (Mustafa 2010). The interprovincial water issue in Pakistan took a worse turn after the worst floods in the country's history in 2010. Pro-dam groups have increased their demands for construction of more storage on the Indus and that has created further apprehensions in the downstream provinces. Popular protests against the proposed construction of the Kalabagh Dam, in particular, are quite common in Sindh and also in the Khyber-Pakhtunkhwa.

The Kalabagh Dam was first conceived in 1953 to span the Indus River in Mianwali District of the Punjab at a point bordering Khyber-Pakhtunkhwa province. Its purpose was to store water for irrigation purposes only. In the 1970s and 1980s, the design of the project was changed to make it a multipurpose large dam of 260 ft elevation. The proposed dam would submerge 35,000 acres of land, generate 3,600 MW of hydropower, store 6.7 MAF of water for flood control, and provide 12.8 MAF water for irrigation. However, the Kalabagh Dam project has remained stalled amidst a fierce controversy ever since former President Pervez Musharraf announced its construction in 2004. Though Punjab is in full support of the dam's construction, major opposition comes from the Sindh and Khyber-Pakhtunkhwa provincial governments, each with its own list of grievances against Punjab. These two downstream provinces are particularly fearful that the project will bring them only water deprivation, ecosystem degradation, and population displacement.

Besides providing large water storage for flood control, the other vital benefits associated with the construction of Kalabagh Dam are irrigation water and power generation. The dam would help to irrigate 800,000 acres of agricultural land and would add at least 2,400 MW of hydropower. However, politically, the dam has been a non-starter in Pakistan as its benefits are viewed as accruing to the Punjab, at the expense of Sindh and the Khyber-Pakhtunkhwa provinces.

While the Indus Waters Treaty has brought a formal river-sharing agreement between two riparian states, India's provinces, especially the Punjab, Haryana, and Rajasthan, continue to fight over their rightful share of the waters of the Ravi, Beas, and Sutlej rivers. The same is also the case among the provinces of Pakistan, who regularly contend with one another over the waters of the Indus, Jhelum, and Chenab. The Indian-administered Jammu and Kashmir state has even started opposing the 1960 Treaty and demanding its review. Domestic political resistance in both countries has stopped the implementation of extremely critical water development projects in the basin. The half-completed Sutlej Yamuna Link Canal in India and the long-planned Kalabagh Dam projects are clear testimony of the costs implicit in the lack of working cooperation over water sharing, not only between but also within the riparian countries.

Harnessing the Indus River is one of the crucial issues for improving human development and contributing to regional peace and security. Paving the way toward enhanced regional cooperation on the Indus River requires comprehensive efforts from many stakeholders. The extreme delicate bilateral relationship especially between Pakistan and India is due to the trust deficit originated since the decolonization of the Indian subcontinent. Uncertainties on the interpretation of the Indus Waters Treaty on infrastructure development in the western rivers have become a major source of friction between India and Pakistan. On the other hand, the water scarcity in the face of rapidly increasing demand has been the source of ethnic mobilization within India and Pakistan. The lack of riparian noncooperation is not only adversely affecting the best possible use of water resources of Indus River system in India and Pakistan, the other riparian countries, particularly Afghanistan are also suffering.

3.4 Afghanistan's Planned Dams on the Kabul River

The challenge in the Indus and its tributaries shared between Afghanistan and Pakistan is represented by the lack of transboundary agreement where the shared waters are crucial to the livelihoods of millions of people using water for vital human needs, agricultural development, and power generation. Kabul River, a tributary of the Indus, supports seven millions Afghani population and also the capital city, Kabul. The Kabul River Basin encompasses around 12 % of Afghanistan's territory and accounts for about 26 % of Afghanistan's total annual river flow. The country has enough of water, but due to lack of infrastructure, the water supply is not available for human consumption. Moreover, the economic recovery of Afghanistan requires investment in irrigation and hydropower. For peace and stability to return to Afghanistan, it is crucial that the water sector development is absolutely essential and unavoidable (Kugelman et al. 2011).

There are already four hydropower dams (the Mahipar, Naghlu, Sarobi I, and Sarobi II) in the Kabul River basin in Afghanistan. In the absence of any agreement between Afghanistan and Pakistan for the sharing of the Kabul River, Pakistan does not pay for flow control structures or management of the river water flow, while Afghanistan also does not share river flow data (Hanasz 2011). In 2010, Afghanistan has come out with a plan to build 13 more dams on the Kabul River with a total water storage capacity of 4.7 million acre-feet. Four of these projects will be built in Panjshir sub-basin: Totumdara (200 MW) project, Barak (100 MW) project, Panjshir I (100 MW) project, and the Baghdara (210 MW) project. In the Logar Upper Kabul sub-basin on the Kabul River, four more hydropower dams are proposed: Haijana (72 MW) project, Kajab (15 MW) project, Tangi Wadag (56 MW) project, and Gat (86 MW) project. Other five proposed dams are in the Lower Kabul sub-basin: Sarobi II (210 MW) project, Laghman A (1,251 MW) project, Konar (A) (366 MW), Konar (B) (81 MW), and Kama project (11.5 MW) (Mustafa 2011). Besides, supporting increasing energy of the country, the cascade of dams will be able to irrigate approximately 184,000 ha of land in Afghanistan (The World Bank 2010).

However, the water storage with the help of these dams can possibly help flood control in Pakistan as well. The excess hydropower from Afghanistan can be also a cheaper and better energy source for Pakistan. But, in the absence of any water sharing agreement and due to bilateral mistrust, Pakistan fears that these proposed dams will reduce its water supply from the river system. In addition, Afghanistan's plan to build these dams has been fuelling political sensitivity around the issue as well.

Media reports (Mustafa 2011) regarding Indian support to Afghanistan for these projects in the Kabul River has further added to Pakistan's worries. Pakistan fears that India's engagement in developing infrastructure on the Kabul River will provide its main adversary another strategic advantage. Though India has not officially confirmed its involvement in the proposed dam building, however, India is very closely and heavily involved in Afghanistan's post-conflict reconstruction projects, and within that framework, water development is a very crucial part.

It is not that Pakistan is only worried about Afghanistan's planned cascade of dams on the Kabul River, similarly India is also getting tense about China's dam projects on the upstream of Indus River. Even, the report of a small-sized Chinese dam near Demchok, Ladakh, on the Indus has ruffled Indians, and they suspect that the China might soon go for building dams on the Sutlej upstream (Sering 2010). In order to achieve basin-based cooperation on promoting food security, enhancing livelihoods, and developing infrastructure in the Indus River system, confidence building between India, Pakistan, China, and Afghanistan can be seen as prerequisite.

3.5 Shared Benefits of Indus Cooperation: A Lost Opportunity

Since 1960, an agreement over the sharing of Indus water between India and Pakistan has been reached. This 1960 Treaty has failed to positively influence the overall bilateral relationship between India and Pakistan. The long-run positive effect of the Indus Waters Treaty is quite disappointing. It has failed to reduce suspicion, mistrust, and uncertainty. Many of the water development projects have remained confined to planning papers only, restricting the possibility of efficient use of scarce water resources. Recurring bilateral disputes over the water projects have raised serious doubts about the possibility of developing the water resources of the river system in a mutually beneficial manner (Swain 2004).

The water scarcity situation is gradually worsening in the basin. Any prospects of integrated water resource development, which is the only long-term answer to basin's growing thirst, have not yet materialized basically due to lack of trust and confidence between two major riparian countries, India and Pakistan. Some of the important water development projects have failed to materialize, not only in India and Pakistan but also in Afghanistan due to bilateral mistrusts and internal politics. For the appropriate and competent management of Indus systems, it is necessary to establish an effective and independent river basin organization, involving all the four riparian sates, which will have the capability of taking decisions on its own and will remain out of the political control of any national government. Under an integrated program of basin development, water projects can be situated at optimum locations, notwithstanding geographic divisions along political lines.

References

Ahmad S (2012) Water insecurity: a threat for Pakistan and India. Atlantic Council Issuebrief

Ahmed S, Naveed A, Shakoor A, Kaleem Ullah M (2007) Meeting future food demands of Pakistan under scarce water situation. Paper no. 667, Pakistan engineering congress, 70th annual session proceedings, Lahore

Arnoldy B (2010) The other Kashmir problem: India and Pakistan tussle over water. Christian Science Monitor, 11 Aug

Bhutta Z (2011) Water wars: India planning 155 Hydel projects on Pakistan's Rivers. The Express Tribune, 15 Nov

Briscoe J, Qamar U (2006) Pakistan's water economy: running dry. The World Bank, Islamabad

D'Souza R (2006) Water in British India: the making of a 'Colonial Hydrology'. Hist Compass 4(4):621–628

Desai BH, Sidhu BK (2014) The Kishanganga final award: is the Indus water treaty at the cross-road. Econ Polit Wkly XLIX(7):10–12

Dhaliwal S (2014) Decades on, SYL canal issue hangs fire. The Tribune, 16 Jul

Earle A, Cascao A, Hansson S, Jägerskog A, Swain A, Öjendal J (2015) Transboundary water management and the climate change debate. Routledge, London

Hanasz P (2011) The politics of water security in the Kabul river basin. Future directions, 10 Nov

Kugelman M, Alam, AR, Bakshi G (2011) Peace through water. Foreign policy, 2 Dec

Laghari AN, Vanham D, Rauch W (2011) The Indus basin in the framework of current and future water resource management. Hydrol Earth Syst Sci Discuss 8:2263–2288

Mirza, AB (2013) Performance of dispute resolution mechanism of the Indus water treaty. Pakistan Institute of Legislative Development and Transparency Background Paper, Islamabad

Misra A (2010) India-Pakistan: coming to terms. Palgrave MacMillan, New York

Mustafa D (2010) Hydropolitics in Pakistan's Indus basin. United States Institute of Peace, special report 261, Nov

Mustafa K (2011) India to help Afghanistan build 12 dams on Kabul river. The News, 12 May

Polgreen L, Tavernise S (2010) Water dispute increases India-Pakistan tension. The New York Times, 20 Jul

Rehman H, Ahmed K (2005) Indus basin river system-flood and flood mitigation. Federal Flood Commission, Ministry of Water and Power, Islamabad

Sering SH (2010) China builds dam on Indus near Ladakh. J Def Stud 4(2):136–139

Sumbal MA (2014) Water scarcity heightens tensions between India and Pakistan. Public Radio International, 8 Aug 8

Swain A (1998) Fight for the last drop: inter-state river disputes in India. Contemp South Asia 7(2):167–180

Swain A (2004) Managing water conflict: Asia, Africa and the Middle East. Routledge, London

Swain A (2009) The Indus II and Siachen peace park: pushing the India-Pakistan peace process forward. Round Table 98(404):569–582

The World Bank (2010) Scoping strategic options for development of the Kabul river basin. The World Bank, Sustainable Development Department, South Asia Region, Washington, DC

Uprety K, Salman MA (2011) Legal aspect of sharing and management of transboundary waters in south Asia: preventing conflicts and promoting cooperation. Hydrol Sci J 56(4):641–661

Wani R (2012) J&K government admits to having started work on Wullar Barrage. The Tehelka, 11 Oct

Wheeler W (2011) India and Pakistan at odds over shrinking Indus river. National Geographic, 13 Oct

Zawahri NA (2008) Capturing the nature of cooperation, unstable cooperation and conflict over international rivers: the story of the Indus, Yarmouk, Euphrates and Tigris rivers. Int J Glob Environ Issues 8(3):286–310

Zawahri NA (2009) India, Pakistan and cooperation along the Indus river system. Water Policy 11(1):1–20

Chapter 4
The Ebb and Flow of Water Conflicts: A Case Study of India and Pakistan

Kristina Roic, Dustin Garrick, and Manzoor Qadir

Abstract A growing body of evidence suggests that domestic water conflicts are not only more prevalent and violent than water conflicts at the international level, they can also have regional and international implications. Using India and Pakistan as a case study, this chapter explores how water conflicts within these two countries affect water relations between them. The chapter uses two forms of research. First, it employs event databases to provide a general overview of the frequency and intensity of water conflict and cooperation both between and within India and Pakistan from 1948 to 2014. Second, it draws on expert perspectives to provide more context and analysis of how water conflicts at these two scales—domestic and international—interact. The chapter concludes that water conflicts within India are largely self-contained and have no bearing on its water relations with Pakistan, whereas water conflicts within Pakistan are closely tied to India's actions upstream and therefore have a tendency to irritate water relations between them internationally.

Keywords Indus basin • India • Pakistan • Database • Water conflict • Domestic • International

4.1 Introduction

In today's inter-connected world, the line between domestic and international events can often be blurred. Just as the aftershocks of global events can influence instability domestically, so too can domestic instability impact international events. This is especially true when it comes to water.

K. Roic (✉) • D. Garrick
McMaster University, Hamilton, ON, Canada
e-mail: roick@mcmaster.ca; dgarrick@mcmaster.ca

M. Qadir
United Nations University Institute for Water, Environment and Health, Hamilton, ON, Canada
e-mail: Manzoor.Qadir@unu.edu

Z. Adeel, R.G. Wirsing (eds.), *Imagining Industan*, Water Security in a New World, DOI 10.1007/978-3-319-32845-4_4

Part of the problem is that rivers do not match political boundaries. Some 276 transboundary river basins are shared by two or more countries: 64 in Africa, 60 in Asia, 68 in Europe, 46 in North America, and 38 in South America (UN-Water 2013). As parts of the world gradually dry up (OECD 2012), experts predict this will translate into heightened competition between water-sharing nations (Chellaney 2011, 2013), between neighboring states and provinces, and between cities and rural areas alike (Gleick and Heberger 2014).

Although significant attention has been devoted to the potential of "water wars" between water-sharing nations, a growing body of evidence suggests that water conflicts[1] at the domestic level are a bigger cause for concern. In the 2014 volume of the *The World's Water*, Peter Gleick and Matthew Heberger report that an alarming proportion of water disputes are occurring domestically, with a risk of escalating to the international level:

> Many are small in scale, involving local violence over water allocations and use or violence over local development decisions that affect environmental and economic conditions at the community scale. But more and more of the reported cases have their roots in water scarcity and competition for a fixed resource that is reaching peak limits…some of this violence has been, and will remain, local. But some may spill over into the international arena.

Similarly, a 2013 United States Institute for Peace report concludes that "intrastate conflicts can cause more damage and violence than interstate water disputes…[yet] intrastate water disputes rarely gain the level of international attention that interstate conflicts do." And the World Economic Forum's *2015 Global Risks* report singles out water as the top risk for the coming decade, with domestic water conflicts as the highest risk within that category. The report projects that two-thirds of the world's population will experience water stress conditions by 2025, making it clear that water must be considered in peace and security programming.

Few regions in the world offer more examples of bitter tensions over water at the local, state, and international levels than the Indus basin and its four riparian states—Afghanistan, China, India, and Pakistan. The question is do the water conflicts at these different scales happen in isolation or do they interact?

This chapter is part of a larger volume that asks readers to imagine "Industan," a reconfigured Indus basin in which Afghanistan, China, India, and Pakistan manage water resources together, in a joint and integrated way. But the potential for transboundary water cooperation cannot be explored without an analysis of water conflicts. Using India and Pakistan as a case study, this chapter examines how domestic water conflicts within these two countries affect water relations between them. Although Afghanistan and China are important stakeholders in the Indus basin, India and Pakistan are the focus due to their disproportionate dependence on the basin in terms of water use,[2] their long-standing history of water conflict, and the high frequency and intensity of domestic water conflict within both countries.

The chapter proceeds as follows: first, it employs two event databases—Aaron Wolf's International Water Events Database and Peter Gleick's Water Conflict

[1] In this chapter, water conflict refers to a wide spectrum of both nonviolent and violent activities, including verbal hostility, protests, riots, and the use of water as a weapon of war.

[2] For a detailed breakdown of India's and Pakistan's share of the Indus basin, see Sect. 1.

Chronology—to provide a general overview of the frequency and intensity of water conflict and cooperation between and within India and Pakistan from 1948 to 2014. Next, the chapter draws on select interviews with water experts from India, Pakistan, and other countries to provide more context and analysis. The aim of the interviews is threefold: to assess the water situation within India and Pakistan, to identify the main drivers of domestic water conflicts in both countries, and to analyze the international implications of domestic water conflicts within India and Pakistan.

4.2 Methodology

The chapter's methodology consists of two parts. The first is a general overview of the frequency and intensity of water conflict and cooperation between and within India and Pakistan from 1948 to 2014, using Wolf's International Water Events Database and Gleick's Water Conflict Chronology.

As Table 4.1 demonstrates, the two databases differ in focus, size, and scope. Wolf's database (retrieved from: http://ocid.nacse.org/tfdd/internationalEvents.php) documents more than 7129 international events of water conflict and cooperation between 1948 and 2008. It provides an event summary, names of basins and countries involved, date of event, category of event (e.g., irrigation, navigation, etc.), source of information, and a *BAR (Basins at Risk) Scale* rating (retrieved from http://www.transboundarywaters.orst.edu/database/event_bar_scale.html), which measures the intensity of events, from least to most cooperative. For the purpose of this chapter, results from Wolf's database were used to provide a general overview of the frequency and intensity of water conflict and cooperation between India and Pakistan at the *international* level.

By contrast, Gleick's database (retrieved from: http://worldwater.org/water-conflict/) tracks 343 events of water conflict at the international and subnational level from 3000 BC to 2014. The database provides an event summary, names of parties and countries involved, date of event, category of event (e.g., political tool,

Table 4.1 Comparison of Aaron Wolf's and Peter Gleick's databases

International water events database (Aaron Wolf)	Water conflict chronology (Peter Gleick)
Incidents of water conflict and cooperation	Incidents of water conflict
International	International and domestic
1948–2008	3000 B.C.–2014
Searchable by basin	Searchable by basin and country
7129 entries	343 entries
BAR Intensity Scale (−7 to 7) measures intensity of event, from least to most cooperative	No intensity scale
Event categories: border issues, flood control/relief, hydropower/electricity, infrastructure/development, irrigation, joint management, navigation, technical cooperation, water quality, water quantity	Events categories: control of water resources, military tool, political tool, terrorism, military target, and development dispute

terrorism), and source of information. For the purpose of this chapter, results from Gleick's database were used to provide a general overview of the frequency of *domestic* water conflicts within India and Pakistan. Gleick's database does not track water cooperation or measure the intensity of events; therefore, this information was not available at the domestic level.

Since Wolf's database is only searchable by basin, the search was conducted using the keyword "Indus." Gleick's database is searchable by basin and country; therefore, this search was performed using the following keywords: "Indus," "India," and "Pakistan." In both cases, only events that mentioned India and/or Pakistan were included in the results. For example, water disputes between China and Tibet in the Indus basin were considered irrelevant; therefore, they were not included. The temporal scale 1948–2014 was chosen to capture as many water events as possible from the 1947 Partition, which led to the formation of India and Pakistan, to present day.

The second part of the methodology consists of select interviews with water experts from India, Pakistan, and other countries. In total, 11 experts were interviewed, including government officials, academics, engineers, journalists, and members of civil society. The purpose of the interviews was to assess the water situation in India and Pakistan, identify the main drivers of domestic water conflicts in both countries, and analyze how water conflicts in India and Pakistan affect water relations between them. The interviewees were asked questions such as the following: What are the main triggers of water conflicts in Pakistan and India? Under what conditions do domestic water issues become a matter of international security? How do domestic water conflicts in India and Pakistan affect water relations between them? Due to the semiformal structure of the interviews, there was some variation in questions.

4.3 Limitation of Event Databases

Before providing an overview of the database results, it is important to acknowledge some of the limitations of event databases. First they are predominantly based on news reporting.[3] This is problematic as news reports are based on a causal judgment made by a journalist who determines whether an event is newsworthy or not (Bernauer et al. 2012). Coverage can also vary depending on a journalist's availability. In countries with limited freedom of the press, journalists may even refrain from reporting on certain events in fear of their safety. In short, there are a number of factors that can render news reporting inconsistent and, therefore, not truly representative of the events.

In addition, the coding of events in a database is based on a judgment imposed by the researcher—another subjective layer—which can lead to what Bernauer et al. (2012) call "false positives or false negatives, i.e., coding a non-water event as

[3] Wolf's database is based on historical documents, in addition to news reports.

a water-related event and vice versa." For example, when a water tanker in Pakistan is attacked, it registers as an event of water conflict in Gleick's database, but is the issue water or ethnic or religious tension? As Gleick himself admits, it is not always clear:

> Disputes over control of water resources may reflect either political power disputes or disagreements over approaches to economic development, or both. We believe this is inevitable and even desirable – international security is not a clean, precise field of study and analysis (WCC website).

Some of these limitations will be evident in the database results below. Despite these shortcomings, news reports are still the largest and most accessible source of information on water-related events, and event databases are the most convenient tool for researchers trying to access them. The key is not to use event databases as the *only* strand of evidence. As this chapter demonstrates, the utility of event databases is best served in combination with other forms of research, such as interviews and a review of primary and secondary documents.

4.4 Database Results

This section provides a general overview of the frequency and intensity of water conflict and cooperation between and within India and Pakistan from 1948 to 2014.

4.4.1 International Results

As indicated in Table 4.2, Wolf's database reports 246 events of water conflict and cooperation between India and Pakistan from 1948 to 2014. Of that number, 54 % (133) have been cooperative in nature and 46 % (113) have been conflictive—a difference of 20 events.

Figure 4.1 reveals a clear trend: the majority of conflictive and cooperative water events between India and Pakistan have occurred during two periods, the 1950s and the 2000s. In both instances, the events have been related to water infrastructure development, particularly damming for irrigation and hydroelectricity.

Table 4.2 Results from Wolf's and Gleick's databases for water conflict and cooperation at the international and domestic level from 1948 to 2014

Database	International conflict	International cooperation	Domestic conflict	Domestic cooperation	Total results
Gleick	n/a	n/a	23 total	n/a	23
			7 Pakistan (30 %)		
			16 India (70 %)		
Wolf	113 (46 %)	133 (54 %)	n/a	n/a	246

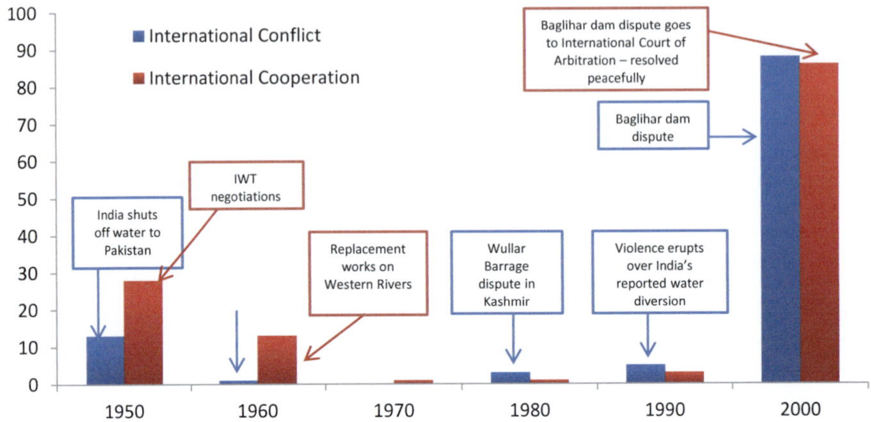

Fig. 4.1 Frequency of international water conflict and cooperation between India and Pakistan from 1950 to 2009 by decade (The *x-axis* represents time. The *y-axis* represents the number of conflictive and cooperative events)

When the political boundaries were drawn in 1947, the majority of the Indus waters were left in Pakistan, but the headwaters remained in India or Kashmir (Collier's 1951). As Condon et al. (2014) state, this division of the river system "set the scene for conflict." The most prominent incident took place in March 1948 when India shut off water flows into Pakistan following the expiration of the "Standstill Agreement,"[4] which was negotiated to allow discharges from headworks in India to continue to flow into Pakistan for a set period of time (Wolf and Newton 2008). According to Wolf's database, this dispute was followed by a period of relative cooperation. Throughout the 1950s, India and Pakistan were engaged in a number of negotiations and agreements in the lead up to the signing of 1960 Indus Waters Treaty (IWT), a water-sharing agreement brokered by the World Bank. Cooperation resumed in the 1960s over the construction of replacement works on the Western Rivers, agreed to under the Indus Basin Project (Wolf and Newton 2008). The only exception was a minor dispute in 1965 over the construction of the Mangla Dam in Kashmir, which resulted in a revolt by Kashmiri workers (International Water Events Database).

Wolf's database reports only one water event between 1967 and 1985, despite the fact that India and Pakistan were at war in 1965 and again in 1971. Although water was not the main cause of these wars, there is arguably no decade in which water between India and Pakistan has not been an issue. Thus, the lack of data during this period cannot be considered a true representation of events.

From the late 1980s onward, water conflicts between India and Pakistan began to escalate. In one incident in 1986, Pakistan objected to India's planned storage

[4] For more information on the "Standstill Agreement," see Wolf and Newton (2008). "Case Study of Transboundary Dispute Resolution: The Indus Water Treaty." University of Oregon. Retrieved from: http://www.transboundarywaters.orst.edu/research/case_studies/Indus_New.htm.

capacity for the Wullar Barrage on Wullar Lake, India's largest freshwater lake in the Kashmir Valley. The project was halted in 1987 due to immense pressure from Pakistan. Through much of the 1990s, water conflicts were triggered by reports that India was planning to divert water from the Neelum River to the Wullar Barrage—an act that would negatively impact Pakistan's irrigation system. In 1997, there was a dramatic escalation in tension over water along the Line of Control in Kashmir, with exchanges of fire by both sides. Five people were injured and 11 Indian troops were killed (International Water Events Database).

The 2000s were marked by a fairly even mix of conflict and cooperation. This was the most active period in terms of events reported, with 36 conflictive events and 35 cooperative events in 2005. The spike in activity was precipitated by India's acceleration of damming for irrigation and hydroelectricity on the Western Rivers. During this period, Pakistan objected to India's storage parameters for two major dams in particular—Baglihar on the Chenab River and Kishanganga on the Jhelum River—claiming they were in violation of the IWT (International Water Events Database; Akhter 2013). In 2005, the Baglihar dispute was referred to the International Court of Arbitration—the first time a dispute had escalated to that level since the signing of the IWT. Pakistan feared that "greater live storage in India could result in flow manipulation, jeopardizing its agricultural production downstream" (Condon et al. 2014). In the end, the World Bank neutral expert—Raymond Lafitte, a Swiss professor of civil engineering—determined that, with some modifications, India's spillways were in accordance with international best practices in dam design (Khattak 2008; Akhter 2013). As India and Pakistan grappled with these issues during the 2000s, their actions translated into in a fairly equal mix of conflictive and cooperative events.

Figure 4.2 provides a general overview of the intensity of water events—conflictive and cooperative—at the international level using Wolf's *Bar Scale* rating (retrieved from http://www.transboundarywaters.orst.edu/database/event_bar_scale.html). The rating ranges from −7 to +7, with −7 denoting the most conflictive events, 0 denoting neutral or nonsignificant events, and +7 denoting the most cooperative events (Yoffe and Larson 2001). Events that scored 0 were not included in the results because they were neither conflictive nor cooperative (there were six such events in Wolf's database). To present the intensity of water conflict and cooperation over time, the scores for each year were added and divided by the number of events to determine the annual mean.

As Fig. 4.2 illustrates, the majority of water conflicts between India and Pakistan scored between −1 ("mild verbal expressions displaying discord in interaction") and −2 ("strong verbal expressions displaying hostility in interaction"). These results imply that water conflicts between India and Pakistan have been largely nonviolent and limited to political and rhetorical statements. Regarding cooperation, most of the events between India and Pakistan scored between 1 ("mild verbal support") and 3 ("cultural or scientific agreement or support—nonstrategic"). These results suggest that water cooperation at the international level has typically been expressed through diplomatic and political statements.

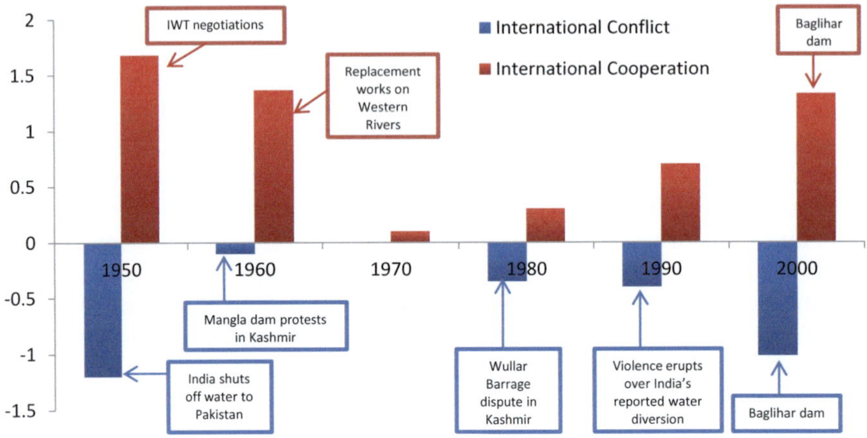

Fig. 4.2 Decennial mean for intensity of international water conflict and cooperation between India and Pakistan from 1950 to 2009 (The *x-axis* represents time. The *y-axis* represents the intensity of conflictive and cooperative events)

4.4.2 *Domestic Results*

As shown in Table 4.2, Gleick's database reports 23 events of water conflict at the domestic level. Of these, 7 % (16) have occurred in India, while 30 % (7) have occurred in Pakistan. Although the sample of data was too small to determine trends—Gleick's database does not report any domestic conflicts in India and Pakistan from 1948 to 1990—we were still able to glean some insights from the results.

In India, the majority of water conflicts captured by Gleick's database have been violent in nature. For example, in December 1999, three people died and 20 were injured in Falla, Gujarat, when the police opened fire on 300 people protesting against the state government's decision to reserve water from the nearby Kankavati Dam for Jamnagar—a neighboring town. The dam had become the only source of water for about 60 villages near Falla, and shortages and overdraft of groundwater contributed to a water crisis (Water Conflict Chronology). In late 2004, four people were killed and more than 30 were injured in ongoing protests by farmers over allocations of water from the Indira Gandhi Irrigation Canal in Sri Ganganagar district, which borders Pakistan. Authorities imposed curfews on the towns of Gharsana, Raola, and Anoopgarh (Water Conflict Chronology). In early 2009, a family in Madhya Pradesh state was killed by a small mob for illegally drawing water from a municipal pipe amidst a terrible drought in the region. Indian media reported more than 50 violent clashes in the state capital Bhopal during May 2009 alone (Water Conflict Chronology). During the summer of 2012, scuffles and protests broke out around New Delhi as residents surrounded water delivery trucks and fought over water. The summer was the hottest in 33 years, leading to extensive energy and water shortages (Water Conflict Chronology).

The number of water conflicts captured by Gleick's database in Pakistan is lower, but even more violent in nature. For example, in 2001, severe water shortages caused by a long-term drought led to riots, four bombs in Karachi, 1 death, 12 injuries, and 30 arrests. Some groups accused the government of favoring the populous Punjab province over Sindh province in water distribution (Water Conflict Chronology). In 2004, military action aimed at Islamic terrorists, including Al Qaeda and the Islamic Movement of Uzbekistan, led to damage and destruction of homes, schools, and water wells. In 2008, the Taliban threatened to blow up Warsak Dam, the main water supply for Peshawar, during a government offensive in the region. During a separate incident in 2010, more than 100 people were killed and scores injured following 2 weeks of tribal fighting over irrigation water in the Kurram region of Pakistan, near the Afghanistan border (Water Conflict Chronology).

Collectively, these events indicate that domestic water conflicts within India and Pakistan have been significantly more violent than water conflicts between them. In India, the majority of water conflicts tend to be centered around access to water for drinking and irrigation. Similar problems are evident in Pakistan, except that water there seems to be most closely linked to terrorist activity, adding an additional layer of complexity to the conflicts.

Despite these general insights, our own rudimentary review of news reports suggests that water conflicts in India and Pakistan are far more prevalent and widespread than Gleick's database results suggest (Atlantic Monthly 2013; Foreign Policy 2015; The Nation 2015). A 2013 report by the United States Institute of Peace even concluded that water disputes in Pakistan "chronic," asserting that "if the current path of water management continues, Pakistan may see the angry protests and isolated acts of violence turn into a larger, more organized form of conflict" (Mustafa et al. 2013). As a result, Gleick's database results are not an accurate representation of domestic water events and require the use of additional sources. The sections below confirm this conclusion and fill the gaps with more context and analysis by expert perspectives from India, Pakistan, and other countries.

4.5 India: Water Conflicts

If one was to look at a map of all the major rivers and tributaries in India, they would see a confluence of blue lines. India has not one but multiple river systems, including the Indus system in the northwest, the Brahmaputra system in the east, and the Cauvery system in the south. But what appears to be a situation of water abundance on paper is not the case in reality.

In recent years, India's population has swelled to 1.29 billion. The National Bureau of Research for South Asia (2013) forecasts that if India continues to grow at its current pace, it will outnumber China as the world's most populous country by

2028. The strains on water are already starting to show. The World Bank (2012) reports that India is the:

> number one user of groundwater in the world. It uses an estimated 230 cubic kilometers of groundwater per year – over a quarter of the global total. More than 60 percent of irrigated agriculture and 85 percent of drinking water supplies are dependent on groundwater.

If these trends continue, the World Bank (2012) predicts that "in 20 years, about 60 percent of all India's aquifers will be in a critical condition."

Expert interviews pointed to three dominant perceptions regarding India's water crisis. The first is that wasteful agricultural practices, propped up by inefficient water policies, are draining the majority of the water:

> Technically on paper, there is enough water. But physical scarcity is starting to creep in. In Punjab, for example, the breadbasket of India, where there are fields and fields of corn, they have completely used up their groundwater. It's an issue of them [farmers] overusing existing resources. (Interview, Raj, Nov. 6, 2015)
>
> I think if you want to talk about why there are such high levels of water stress in these regions – you could talk about population growth, you could talk about economic growth, you could talk about climate change effects, that's all there – but honestly, I think you have to direct a lot of blame to Indian water policies. The fact that, consistently, Indian farmers, who are a very important political constituency, use the most wasteful types of irrigation – flooding irrigation because that's what they've always used and that's what's given to them basically for free. There are ways of using less irrigation water, or using it more efficiently, through drip-irrigation forms, but those are not subsidized. So, as a result, you have situations where you literally flood your fields because you have so much water and a lot of times it gets pumped in at night, even when you don't need to use it. It's incredibly wasteful. And it's ironic because this is what's making the water disappear. (Interview, Kugelman, Oct. 30, 2015)
>
> Unfortunately, water-intensive crops like rice are being grown in semi-arid areas like West Punjab. I would argue that we should switch to coarse grain and shift rice somewhere else. If India focused on making agricultural practices more efficient, the water problem would be solved. Drip-irrigation, more crop per drop – that would be much more sensible. (Interview, Gautam, Nov. 2, 2015)

The second dominant perception is that as more water disappears and is not replenished, people are more willing to act in violent ways to avenge their grievances. This has translated into increased water conflicts at all levels:

> There are so many disputes. It's a nightmare for a scholar to analyze all the water disputes within India. (Interview, Gautam, Nov. 2, 2015)
>
> Water disputes are increasing at every level, from local to inter-state to international. Essentially, water is a local issue because the demands are local. The groundwater is also essentially a local resource, so there are more and more disputes happening at the local level. There are more disputes also at the inter-state level and also at the international level. (Interview, Thakkar, Oct. 27, 2015)

The third dominant perception is that local water conflicts tend to be related to access, whereas interstate conflicts tend to be related to dams:

> In many places, people don't have access to water so they have to share a common resource. For example, in the slums, there are community supplies and this leads to violence between families and communities. Inter-state disputes generally don't end up in violence. They take different forms. There are places where people have come on the streets to protest, but in

> most instances, they become a political issue between two states and then ends up in the courts, at the inter-state tribunals set up to resolve the disputes. But what I have noticed is that the disputes at the inter-state and international level mostly emanate from the agenda for big dams. Generally, as long as there is no agenda for big dams, there are no disputes. (Interview, Thakkar Oct. 27, 2015)
>
> I think access in terms of disputes at the local level are the biggest issue because the majority of Indian farmers are small-holder farmers. The average land holding for a small farmer in India is 0.8 hectares – that's a *tiny* piece of land. And often, they don't have access to water for their small parcel of land. Water is taken up by large landowners, by special interest groups, by commercial farms, which produce sugar cane or other water-intensive crops. So, it's a little guy versus big guy issue. The people who need water the most have the least amount of access to public policy. (Interview, Raj, Nov. 6, 2015)

4.6 Pakistan: Water Conflicts

Unlike India, which can draw from a number of river systems, Pakistan's geography makes its population almost entirely dependent upon the Indus basin for agriculture and drinking water. Of the 237 million people who are supported by the basin, Pakistan accounts for 61 % of the population or 145 million people, in contrast to India, which accounts for 35 % of the population or 83 million people (Laghari et al. 2012).[5] This automatically renders Pakistan's water situation more precarious than that of its upper riparian.

As with India, Pakistan's population is a significant factor. At the time of its formation in 1947, Pakistan had 31 million people. By 1995, it had 140 million (Mahsud-Dornan 2007). Today, it has 191 million people (Pakistan Economic Survey 2014–2015). This rapid population growth—combined with urbanization and increased agricultural and industrial demands—has turned Pakistan into one of the most water-stressed countries in the world, coming dangerously close to being classified as water scarce (Sharjeel 2006, ADB 2013). The United Nation's Food and Agriculture Organization considers water stress to be high if the pressure on national renewable water resources is above 25 %; Pakistan's is at an alarming 74 % (Mustafa et al. 2013).

Expert perspectives offered three dominant perceptions in regard to Pakistan's water crisis. The first is that Pakistan's water problems have less to do with actual physical scarcity of water and more to do with inequitable water policies, corruption, and weak institutions:

> The problem is not 'is there enough water?' There is definitely enough water in the Indus Basin that enters Pakistan. The problem is that it's in the south and the people who need the water aren't getting enough of it. Whereas if you look at the cities, people have grass lawns, the army has huge complexes that it keeps very well-watered, all the cities are well hooked up. So, I don't know if the question is one of 'is there enough water?' I just think

[5] Data on this varies, depending on the source. For example, the Strategic Foresight Group (2011) reports that the Indus basin covers 71 % of Pakistan's territory and provides water for 77 % of the population.

> there needs to be the proper infrastructure and changes in institutions to get the water where it needs to go… I think the reason that it [Pakistan] keeps operating at a sub-optimal level is precisely because some key people will remain in power, even if nothing gets done. Pakistan tends to be jumping from crisis to crisis, and when a crisis occurs it does not affect everyone. There is always an elite that comes out ok, so nothing changes. (Interview, Akhter, July 6, 2015)
>
> The provincial institutions in Pakistan that are meant to mediate water tensions and concerns between these provinces are completely feckless. They are also corrupt and they don't work, which is very different from something like the Indus Waters Treaty, so there is a lot more fragility and vulnerability at the domestic level. (Interview, Kugelman, Oct. 30, 2015)

The second dominant perception is that water conflicts in Pakistan are happening, though opinions diverged to what degree:

> Water is the number one issue for the nation. Pakistan is facing a humanitarian crisis that could lead to a regional war and worse. There is insufficient ground water recharge, but the Pakistani people have predominantly to rely on ground water for their drinking needs. Extremely serious bacterial and heavy metal contamination is becoming irreversible. Child mortality in Pakistan is higher than in sub-Saharan Africa. The cake is getting smaller. There is an increase in fighting over water at the farm-gates and between the provinces. (Interview, Khan, July 30, 2015)
>
> Population is increasing and demand is increasing. When demand is increasing, you see competition and when competition is there, conflict starts. I have observed people shouting, crying, farmers going on hunger strikes, sitting on the road, blocking the road. I have observed in my projects that people are really suffering. (Interview, Rana, June 27, 2015)[6]
>
> Of course there is contention. But these are small skirmishes between one area to another, usually within the same province. It has happened in the Punjab province and it has happened in the Sindh province, but there are no violent disputes resulting in rioting, clashes with police, or terrorist attacks on water infrastructures. (Interview, Memon, August 7, 2015)

The third dominant perception is that the main flashpoint for water conflicts is dams—or the lack thereof:

> The nation has not built a large dam in 41 years. Mangla dam on the Jhelum River was commissioned in 1967. The sole dam on the Indus River was completed in 1974. Not even a replacement reservoir could be created as a 'replacement' of the storage capacity lost in these 41 years while its population has at least tripled. I admit that Pakistan has been callous, ignorant and dis-united on the subject of its water endowment. (Interview, Khan, July 30, 2015)
>
> There are two kinds of thinking in Pakistan right now: should there be big dams or should there not be? This is the debate going on. (Interview, Memon, August 7, 2015)

According to the Asian Development Bank (2013), Pakistan's current storage capacity is limited to a 30-day supply—far below the internationally recommended 1000 days for countries with similar climates. The situation has become so dire that in March 2015, Pakistan's Water and Power Minister Khawaja Asif told media that the construction of dams is a "matter of life and death" (Nayani 2015).

[6]The opinions expressed here are the subject's personal views. They do not represent the views of the ADB or the government of Pakistan.

The experts interviewed held disparate views on why it has been so difficult to get dam projects off the ground. Some claimed that interprovincial politics were the culprit:

> The provincial governments don't trust each other, even though there are water measurement devices and telemetry measuring systems. This is part of the larger dispute in Pakistan – this kind of overly dominant position of Punjab within the federation. This tends to be the case when the federal government happens to be dominated by one province, which in this particular case also happens to be the upstream province. One is not causing the other directly, but they are kind of reinforcing each other. So, you have all these different types of unevenness feeding into each other. (Interview, Akhter, July 6, 2015)

Others argued that interprovincial rivalry is only part of the story:

> The internal faults are there, but in the realm of 'the politics of water' the Indians are terribly complicit. They have stage-managed the anti-dam movements in Pakistan for 40 plus years, and in the shadow of the East Pakistan debacle of 1971, the Punjabi people have allowed the smaller provinces to literally get away with murder. A family that opposed the creation of Pakistan and compelled the father of the nation (Mr. Jinnah) to call a referendum on whether NWFP (now KP Province) will become part of Pakistan – this family remains at the forefront of the anti-dam movement. And Punjab and the national leadership is in a state of paralysis on this vital issue. Some of the provincial leadership seem to be running auction houses where India is the only bidder. India takes full advantage of this induced situation. (Interview, Khan, July 30, 2015)

A Pakistani journalist, who wished to remain anonymous, admitted that her initial skepticism about an Indian anti-dam movement in Pakistan was proven wrong upon further investigation. Her research found that Indian groups are not only funding the blockage of dams in Pakistan, they are also funding the blockage of information:

> Initially, I used to think that these are all conspiracy theories—foreign-funded agencies and foreign-backed groups—but the more time I spend in journalism, the more I realize these things do exist. There are groups that are backing the blockage of this information to our public. There are television channels that refuse to report on it. There are newspapers that refuse to write about it. Their editorial policy simply is "don't talk about water." And it doesn't make any sense because it is such a serious issue. (Interview, August 3, 2007)

One expert confirmed that India "has definitely been doing this and not just in Pakistan, but in Afghanistan and Bangladesh, too" (Interview, Memon, August 7, 2015). However, he did not think it had actually prevented any projects from going forward.

4.7 Interaction Between Domestic and International Water Conflicts

This section examines how domestic water conflicts in India and Pakistan impact water relations between them. The prevailing view among experts who were interviewed is that domestic water conflicts in India are largely self-contained:

> I don't see domestic issues spilling over internationally. First of all, the parts of the country that do share water, which is the north-east primarily and the west, are very stable. The west is politically very stable and it is moving away from an agrarian economy. And the north-east is water abundant right now, though obviously, 20 years from now, that situation could be quite different. Second, domestically, India has a very strong institutional framework. You've got municipal water authorities, you've got city water authorities, you've got district-level authorities, state, central, and so on. Finally, not all of our water is shared. You have a huge chunk of the country – for example Maharashtra or Andhra Pradesh – that doesn't share water with other countries. (Interview, Raj, Nov. 6, 2015)

However, in the case of Pakistan, experts felt that domestic water conflicts could have an impact on water relations with India at the international level:

> In Pakistan, I think these issues could spill over into the international arena because the majority of their water is shared with Afghanistan or India, so it's very difficult for them to look at a domestic issue in isolation of an international issue. Also, Pakistan does not manage its water well. The institutional framework, the WAPDA for example, is a fairly iffy, Indus-active body that has a tendency – as one former Indus Water Commissioner has come out and said – to blame international reasons or a foreign hand, rather than saying 'look, we have a water issue.' They have one of the largest, continuous irrigation networks in the world and yet, we did some rudimentary calculations and we found that over 40 percent of their water is wasted. It's an inefficient sector backed up by inefficient institutions. For those reasons, I think its internal disputes could escalate. (Interview, Raj, Nov. 6, 2015)

Experts unanimously agreed that Pakistan has a strong tendency to blame India for its water crisis:

> It [Pakistan's water crisis] is linked to India's dams in the common public, whether it is in reality or not. This is the easiest way in Pakistan to divert the attention away from the issues. There are, of course, problems among the provinces of Pakistan, but mostly, the opinion it is that India has built a dam and it is creating a problem and the water shortages are due to Indian dams. (Interview, Rana, June 27, 2015)[7]
>
> I think anything that has to do with the Indus will always be Indianized just for the very fact that it's connected to India or part of the common basin. It's one of the inescapable points of unity and creation between the regions, despite the partition. But the more serious questions about water and Pakistan don't have anything to do with India and everything to do with uneven development within Pakistan. (Interview, Akhter, July 6, 2015)

Experts confirmed that this blame displacement carries potentially serious consequences. They explained that in recent years, militant groups in Pakistan have been exploiting the country's internal water insecurity to incite hatred against India:

> We [the Strategic Foresight Group] do a lot of work with our domestic partners in Pakistan and we've seen that water is an issue that is being used right now to radicalize young people against India. It's a recruitment tactic. It's playing into the very worst fears of the very poor, underserved population for whom water is very central to their lives. Terror groups in Pakistan, many of whom function quite freely or have free access to people, will talk to incredibly young people, or children of farmers, for example, who live in extremely rural areas and they will say 'the reason you don't have water is because of India.' There have even been reports in Pakistani and jihadi papers that say that India is creating cloud conditions favourable to itself in an effort to prevent Pakistan from having water. It's blatantly

[7] The opinions expressed here are the subject's personal views. They do not represent the views of the ADB or the government of Pakistan.

fictitious, of course, but water is very easy to play upon because it is such an emotive issue that people get really incensed about it. (Interview, Raj, Nov. 6, 2015)

The level of hatred and raw emotions are so high on this side [Pakistan]. The ten idiots who went to shoot-up Mumbai in 2008 were boys from South Punjab where the lands have literally become barren within three decades after the Indians took away the three Eastern rivers (Ravi, Sutlej and Beas) as part of the 1960 IWT. What will happen if the bulk of the nation becomes suicidal? (Interview, Khan, Aug. 1, 2015)

Several experts claimed that these domestic hate campaigns are making India think twice about development plans in the Indus basin:

The Indians have made a very concerted effort to hold back on building more dams in Kashmir so as not to risk the possibility that Pakistan would become unhappy and the LAT [Lashkar-e-Taiba] would get up in arms. The LAT is the group that has been most vociferous in accusing India of stealing water and I think there is a lot of concern that it could ramp up that rhetoric and perhaps even use it as a pretext for some sort of attack on India. But I think the issue is that as India's domestic water situation becomes more precarious, it's not going to have as much of an incentive to hold back from doing things that would violate the Indus Waters Treaty, such as building dams that would actually store more water and stop it from flowing downstream. I would actually argue that the new government in India, which took office last year, is more conservative that the previous ones, and it has said that it's not going to be afraid of provoking Pakistan, so that could lead to a rather ugly situation. (Interview, Kugelman, October 30, 2015)

Unfortunately, because of geopolitical thinking, Pakistan is unhappy of any Indian development projects in the Indus basin, so you will find India revising and making smaller dams. (Interview, Gautam, Nov. 2, 2015)

However, not everyone was persuaded:

The Indian government is watching militant groups and it is being cautious about it, but I don't think it has prevented them from executing a number of projects. I think there were a number of projects that were on paper that people say, 'oh those haven't been executed,' but realistically speaking you can't build 12 dams on three and a half rivers that you share with Pakistan. The one good thing is that there is a mechanism for both countries to refer each other, but India finds that every time they try and build something, it's gone to court where it takes a number of years to resolve. So from that angle, it has been a bit slower and they are looking to diversify because hydropower is their primary interest in building these dams. Run-of-the-river dams are technically allowed by the IWT, but it still becomes a complex bargaining business. So, I don't think it's put a damper on any plans, but I would say that India is proceeding with the dams that it's very intent on. It's a prioritization thing I think. (Interview, Raj, Nov. 6, 2015)

4.8 Conclusion

For two countries who were "born into conflict" (Wolpert 2010), India and Pakistan have a lot in common. Both countries have rapidly growing populations. Both are water stressed, with Pakistan hovering on the verge of scarcity. Both are agrarian economies, with wasteful and inefficient agricultural sectors. Both have depleted groundwater reserves. Both have uneven development, with the breadbasket regions using the lion's share of water. Both are plagued by water mismanagement, inequitable distribution, and inefficient water policies, though India has stronger institutional frameworks than Pakistan. Finally, both struggle with domestic water conflicts at the local and state levels. At the local level, the disputes tend to be related to access, whereas at the state level, they are commonly triggered by dam projects. However, this is where the similarities between India and Pakistan end.

As this chapter demonstrates, water conflicts in India are largely self-contained and have no bearing on its water relations with Pakistan. This is due to the fact that India can draw water from multiple river systems and has a fairly robust institutional framework to deal with such disputes. For Pakistan, the situation is more precarious. Because the country is almost entirely dependent on the Indus basin for its agriculture needs and drinking water, Pakistan's domestic water conflicts are not only closely tied to its water relations with India, they have a direct impact on how the two countries approach water sharing in the Indus basin. Being the lower riparian country, Pakistan is hypersensitive to any activity in the Indus basin and tends to see its domestic water problems closely associated with India's actions upstream. As a result, Pakistan has a strong tendency to blame India for its internal water problems, with militant groups like the LAT retooling its public relations efforts around the water crisis to recruit young people and incite hatred against India.

Several experts claimed that this has had an adverse effect on India's development projects in the Indus basin, pointing to signs that India is slowing down or opting for smaller and less controversial run-of-the-river dams. However, these conclusions are based on anecdotal evidence and require more in-depth surveying in order to be truly representative. The concern, however, is that as water crises in both countries intensify and resources become even further stretched, Pakistani militant groups will ramp up its anti-India efforts as India accelerates its construction of large dams on the Western Rivers in order to meet its demands. In that scenario, Pakistan's domestic water instability could further irritate its water relations with India and fuel other hostilities at the international level.

References

Akhter M (2013) Geopolitics of dam design on the Indus. Econ Pol Wkly 19:24–26Asian Development Bank (2013). Asian Development Outlook 2013: Asia's Energy Challenge, p. 208. Retrieved from: http://www.adb.org/sites/default/files/publication/30205/ado2013-pakistan.pdfBernauer T, Tobias B, Halvard B, Nils PG, TheresaT, Eivind BW, and Gerdis W (2012). "Water-Related Intrastate Conflict and Cooperation (WARICC): A New Event Dataset." International Interactions, 38: 529–545.

Asian Development Bank (2013). Asian development outlook 2013: Asia's energy challenge, p.208. Retrieved from: http://www.adb.org/sites/default/files/publication/30205/ado2013-pakistan.pdf

Bernauer T, Tobias B, Halvard B, Nils PG, Theresa T, Eivind BW, Gerdis W (2012) Water-Related Intrastate Conflict and Cooperation (WARICC): a new event dataset. Int Interact 38:529–545

Chellaney B (2011) Water: Asia's new battleground. Georgetown University Press, Washington

Chellaney B (2013) Water, peace and war: confronting the global water crisis. Rowman & Littlefield, New York

Condon M, Kriens D, Lohani A, Sattar E (2014) Challenge and response in the Indus basin water policy. Water Policy 1(30):1–29

Gleick P, Heberger M (2014) Water and conflict: events, trends, and analysis. In: The world's water, vol 8, Pacific Institute, January 2014

Interview with Akhtar Rana, principal water source specialist for the Asian Development Bank, 27 June 2015

Interview with Anumita Raj, senior programme manager at the strategic foresight Group, 6 Nov 2015

Interview with Himanshu Thakkar, coordinator of South Asia Network on Dams, Rivers and People in India, 27 October, 2015

Interview with Majed Akhter, assistant professor of geography at Indiana University, 6 July 2015

Interview with Michael Kugelman, senior manager of South and Southeast Asia at the Woodrow Wilson International Center for Scholars, 30 October 2015

Interview with Nisar Memon, former Senator of Sindh province and current chairman of the Water Environment Forum in Pakistan, 7 August 2015

Interview with Pradeep Gautam, research fellow at the International Defence Studies and Analyses, 2 November 2015

Interview with Suleman Najib Khan, independent engineering consultant and co-founder of the Water Resource Development Council in Pakistan, 30 July, 1 and 3 August 2015

Interview with anonymous journalist from Pakistan, 3 August 2015

Johnson K (n.d.) Pakistan's energy crunch fuels little but outrage. Foreign policy, 23 January 2015. Retrieved from http://foreignpolicy.com/2015/01/23/pakistans-energy-crunch-fuels-little-but-outrage-sharif-protests-khan/

Khattak AR (2008) World bank neutral expert's determination on baglihar dam: implications for India-Pakistan relations. Pakistan Horizon 61(3):89–108

Laghari AN, Vanham D, Rauch W (2012) The Indus basin in the framework of current and future water resources management. Hydrol Earth Syst Sci 16:1063–1083

Lilienthal D (1951) "Another 'Korea' in the making?" Collier's, 4 August 1951.

Mahsud-Dornan S (2007) Pakistan, population programmes and progress. Ulster Med J 76(3):122–123. Retrieved from http://www.ncbi.nlm.nih.gov/pmc/articles/PMC2075591/

Mustafa D, Akhter M, Nasrallah N (2013) Understanding Pakistan's water-security nexus. United States Institute for Peace, Washington

Nayani A (2013) Pakistan's new big threat isn't terrorism – it's water. Atlantic Monthly, 19 July 2013. Retrieved from http://www.theatlantic.com/international/archive/2013/07/pakistans-new-big-threat-isnt-terrorism-its-water/277970/

Nayani A (2015) Pak minister asks India to honour Indus waters treaty. Zee News, 9 March 2015. Retrieved from: http://zeenews.india.com/news/south-asia/pak-minister-asks-india-to-honour-indus-waters-treaty_1558888.html

OECD (2012) The consequences of inaction in the OECD environmental outlook to 2050. Retrieved from http://www.oecd.org/env/indicators-modelling-outlooks/waterchapterofthe-oecdenvironmentaloutlookto2050theconsequencesofinaction.htm

Pacific Institute's Water Conflict Chronology. Retrieved from http://worldwater.org/water-conflict/

Population, labour force and employment. Chapter 12 in Pakistan economic survey 2014–15. Retrieved from http://www.finance.gov.pk/survey/chapters_15/12_Population.pdf

Sharjeel S (2006) Better management of the Indus basin waters: strategic issues and challenges. World Bank, Islamabad

Strategic Foresight Group (2011) The Indus equation. Retrieved from http://www.strategicforesight.com/publication_pdf/10345110617.pdf

Transboundary Freshwater Dispute Database (TFDD). International water events. University of Oregon. Retrieved from http://www.nbr.org/research/activity.aspx?id=356 http://ocid.nacse.org/tfdd/internationalEvents.php

UN-Water. International Year of Cooperation Facts and Figures (2013). Retrieved from http://www.unwater.org/water-cooperation-2013/watercooperation/facts-and-figures/en/

Wolf AT, Newton JT (2008) Case study of transboundary dispute resolution: the Indus water treaty. University of Oregon. Retrieved from http://www.transboundarywaters.orst.edu/research/case_studies/Indus_New.htm

Wolpert S (2010) India and Pakistan: continued conflict or cooperation? University of California Press, Los Angeles

World Bank (2012) India groundwater: a valuable but diminishing resource. Retrieved from: http://www.worldbank.org/en/news/feature/2012/03/06/india-groundwater-critical-diminishing

World Economic Forum (2015) Global risks 2015. Geneva. Retrieved from http://www.weforum.org/reports/global-risks-report-2015 http://www3.weforum.org/docs/WEF_Global_Risks_2015_Report15.pdf

Yoffe S, Larson K (2001) Chapter 2 Basins at risk: water event database methodology. Transboundary freshwater dispute database website. Retrieved from http://www.transboundarywaters.orst.edu/database/Yoffe%20&%20Larson-Event%20Coding.pdf

Part III
The Potential for Transboundary Cooperation

Chapter 5
The Indus Waters Treaty: Modernizing the Normative Pillars to Build a More Resilient Future

Bjørn-Oliver Magsig

Abstract While the fact that, despite their strained relations, India and Pakistan managed to negotiate the Indus Waters Treaty (IWT) has been widely celebrated as a success, tensions concerning how to share the water resources of the Indus are rising again. The apparent mismatch between analysis and reality is due to the fact that, so far, most scholars have asked the wrong questions when it comes to water security, international law, and the obligation to cooperate. This chapter will introduce a contemporary understanding of water security which goes beyond the narrow state-centered zero-sum game debate and provides a platform for various disciplines to engage in strengthening cooperation over shared waters. This novel lens will then be applied to analyze the legal framework governing the utilization of the Indus. Can the concept of common concern for water security actually be implemented through the IWT? In marrying security studies with international law, potential futures for the legal framework governing the Indus will be illustrated. It is hoped that this chapter will shine a new light on the question whether the IWT is up to the task of providing water security by building a more resilient future for the basin and whether international law has a role to play in bringing about Industan.

Keywords Indus Waters Treaty • International law • Treaty interpretation • Common concern • Duty to cooperate

5.1 Introduction

Very few challenges have the potential to create as much friction between states as the allocation and utilization of freshwater resources which cross international boundaries. The various simmering water-related conflicts around the world – like Egypt's difficult relationship with Ethiopia on the Nile (Abseno 2013; Zhang et al.

B.-O. Magsig (✉)
Helmholtz Centre for Environmental Research, Leipzig, Germany
e-mail: bo@magsig.de

Z. Adeel, R.G. Wirsing (eds.), *Imagining Industan*, Water Security in a New World, DOI 10.1007/978-3-319-32845-4_5

2015), the dispute over a more equitable distribution of water in the Middle East (Fröhlich 2012; Weinthal et al. 2015), or the highly contentious region of Himalayan Asia (Wirsing et al. 2013; Magsig 2015b) – bear witness to the fact that water, the gossamer linking various other security concerns, has acquired an independent status within the security discourse (Wouters et al. 2009). Just as the concept of security has gone through a widening and deepening process, so has the perception of water security (Magsig 2014). The approach of addressing water security merely as a nation's internal affair of securing its access to a continuous supply of freshwater is untenable in today's interrelated world. This is particularly true for the Indus basin, where the co-riparians have struggled to create an environment of fruitful cooperation.

While India, as a regional hegemon and riparian country of the river basins of Brahmaputra, Ganga, and Indus, has recently been in disagreement over freshwater sharing with several of its neighbors (Taenzler et al. 2011), the most prominent dispute is between India and Pakistan over the Indus, which is the bloodstream of Pakistan's economic (textile industry) and food security. Relations between the two neighbors have always been highly tense, since "India and Pakistan were born to conflict" (Wolpert 2010); and thus, it does not come as a surprise to hear even more of the "water wars" propaganda here (Mandhanda 2012). It has been argued that the general perception that India is threatening the survival of Pakistan simply by "cutting off" the waters of the Indus raises the potential of Pakistan using nuclear weapons as a last resort (Brennan 2008).

However, the two countries did manage to negotiate a water agreement despite their strained relations. While the Indus Waters Treaty (Treaty between India and Pakistan Regarding the Use of the Waters of the Indus 1960) has been widely celebrated as a success for putting an end to the water woes between India and Pakistan (Khalid 2004), tensions concerning how to share the water resources of the Indus are rising again. The apparent mismatch between analysis and reality is due to the fact that, so far, most scholars have asked the wrong questions when it comes to water security and international law. In this chapter, it will be demonstrated that a contemporary understanding of water security goes beyond the narrow state-centered zero-sum game debate and provides a platform for various disciplines to engage in strengthening cooperation over shared waters. This novel lens will then be applied to analyze the legal framework governing the utilization of the Indus – of which the Indus Waters Treaty constitutes the cornerstone. In marrying security studies with international law, potential futures for the legal framework governing the Indus will be illustrated. It is hoped that this chapter will shine a new light on the question whether the Indus Waters Treaty is up to the task of providing water security and whether international law has a role to play in bringing about Industan.

5.2 Water Security and International Law

The crosscutting nature of water creates global interdependencies which make solutions to the water scarcity crisis highly complex, as water cooperation cannot be separated from global trends and drivers outside the "water box." Here, international law can and should provide the normative content, as it (1) defines and identifies the legal rights and obligations regarding the use of water and provides the prescriptive parameters for the management of the resource, (2) provides tools for ensuring the continuous integrity of the regime (including dispute prevention and settlement), and (3) allows for modifications of the existing regime, in order to be able to accommodate change (Wouters et al. 2009).

5.2.1 Water Security Through a Legal Lens

When analyzing the emerging global water crisis, one soon realizes that the past may not be an adequate basis from which to make predictions about the possibility of future conflicts over water. This is why, based on a pessimistic neo-Malthusian outlook (Homer-Dixon 1994), the notion of "water wars" got traction – not only in the media (Chellaney 2011). Cornucopians, in contrast, draw a rather optimistic picture of the future, which they build around the argument that the water crisis is a crisis of management rather than one of absolute scarcity (Gleditsch 1998). In their view, it will be resolved by anthropogenic means, like international trade in "vir tual" water, economic development, and investment in new infrastructure (Allan 2002; Barnaby 2009). This passionate water war vs. water peace debate often misses a crucial point: Even if the future international conflicts over water are not likely to lead to fully fledged wars, strengthening cooperation between the riparian states will still help alleviate the water crisis. The fact that every year more than 3.5 million people die because of poor water, sanitation, and hygiene – far more than by all the ongoing wars combined – clearly suggests the need for a wider approach to water security than the narrow military one (Schuster-Wallace et al. 2008). Even if disagreement between riparian states over the allocation and utilization of their transboundary freshwater resources may not always pose a direct military threat, it nonetheless has the potential to destabilize societies in a world which – in some regions – is already highly unstable (Magsig 2014). General security studies also followed this understanding with the inclusion of nonmilitary threats ("widening") and efforts to "deepen" security research (Buzan 1991). With this new approach – called human security – the individual, rather than the state, is being regarded as the chief referent object (von Tigerstrom 2007).

Furthermore, it became obvious that a new strategy addressing the drivers of insecurity by "curing the disease" rather than "fighting the symptoms" was urgently needed (Brock 2011). This is why recent research on collective and sustainable security is trying to pave the way toward a mutual understanding that security can no longer be regarded as a zero-sum game between states – since a contemporary take on the notion unveils its "common" characteristic (Scholtz 2009; Voigt 2009). In combining sustainable security thinking with the more advanced concept of collective security – also being perceived as one of the core purposes of the UN Charter (United Nations 2004) – a promising platform for discourse is finally emerging which is capable of facilitating a meaningful debate about how to address the various security issues the international community is facing. While states are the bedrock of the international system and thus achieving collective security is impossible without being based on various perceptions of states' securities, collective security is operating somewhat "above and beyond" orthodox patterns of international relations – i.e., to add "universal moral obligations" to the table of international negotiations (Orakhelashvili 2011).

Against this backdrop, this chapter follows a broader understanding of security and regards a community to be "water secure when it has sustainable access to freshwater of sufficient quantity and quality, or to the benefits derived therefrom; and the ability to minimize water-related risk and its various repercussions to an acceptable level – without compromising the supporting ecosystems" (Magsig 2015a). This definition draws from both the widening and deepening processes of the general security debate while, at the same time, acknowledging the complexity of the global water crisis. Accordingly, the definition has several advantages over previous (mostly more restrictive) ones. Firstly, by focusing on "communities," it is scalable to the level one wants to look at water security – local, national, regional, or even global. It also acknowledges the fact that in water resource management, the overlapping of several levels of governance is the rule rather than the exception. Secondly, by including the "benefits derived" from access to freshwater and the repercussions of water-related risks, the true complexity of the water crisis is being pulled into play. Not only are we looking at access to and threats from the resource water but also the opportunities and issues linked (directly or indirectly) to it. Here, the concepts of virtual water and benefit sharing come to mind (Wouters and Moynihan 2013). Finally, by entailing undetermined parameters like "sufficient quantity and quality of freshwater" and "acceptable level of water-related risk and repercussions," the definition provides the respective community with considerable room to maneuver concerning the implementation of the concept of water security – geared to its own needs, capacities, and preferences. The relative vagueness of the concept guarantees its resilience as well as global applicability while, at the same time, it avoids becoming arbitrary (Magsig 2015a). Rather than being a somewhat constricting stipulation of the term "water security," it aims at providing a platform for stimulating discourse.

Yet, in order to be able to analyze international legal regimes, this definition needs to be fleshed out further. Earlier work has developed an analytical framework for examining international law through a water security lens by focusing on issues of (1) availability. (2) access, (3) adaptability, and (4) ambit (Magsig 2009). Issues of availability relate to concerns of water quality as well as quantity. This facet deals primarily with the actual management of the resource – including its control and sustainable protection. This includes the need to maintain the natural integrity of the freshwater resource by calling for environmental flows (Forslund et al. 2009). The element of access is central to the water security debate, as it deals with the issues revolving around the right to utilize a shared water resource. Given the complexity of cooperation over water resources, access covers a broad spectrum of concerns across the growing range and number of users and uses with regard to matters of (re) allocation. Here, the principle of equitable and reasonable utilization, the cornerstone of international water law which has reached customary status, is key to the process (Wouters et al. 2005). It determines the right of a state to use the waters of an international watercourse in two distinct ways: (1) by establishing the objective to be achieved, which then specifies the lawfulness of the new (or changed) utilization of an international watercourse, and (2) by incorporating an operational function, since it requires all relevant factors and circumstances to be taken into account when determining what exactly qualifies as an equitable and reasonable use (Rieu-Clarke et al. 2012). In order to support the obligation to weigh and balance all of the stakeholders' interests, dispute prevention and settlement mechanisms are of vital importance (Salman 2006). As, in most cases, the key factor of transboundary water cooperation is not absolute water scarcity, but rather the resilience of the institutions which govern the shared resource, a legal regime for transboundary watercourses has to include flexibility and ensure adaptability to address changing conditions – while still providing for some level of predictability (Magsig 2014). This element deals with the various uncertain variables – e.g., impacts of global environmental change, population growth, and economic development – which influence transboundary water cooperation considerably. However, most freshwater agreements are rather rigid instruments, as they can only be modified according to their own terms or by mutual agreement. Hence, if a treaty lacks flexible tools and water stress soars, disputes over the shared resource are likely to intensify in cases where one party to the agreement may find it difficult to reduce its consumption in order to comply with its legal obligations. If the water stress causes asymmetric harm, the harmed state may be eager to terminate the agreement, while the co-riparian may find it beneficial to stick with it. In this regard, the International Court of Justice (ICJ) concluded in its Gabčíkovo-Nagymaros judgment that "[…] the stability of treaty relations requires that the plea of fundamental change of circumstances be applied only in exceptional cases" (ICJ 1997). The ICJ further noted that new developments or changing conditions should be dealt with on the level of implementation of the treaty, not by simply terminating it. However, several studies come to the conclusion that many states will have to renegotiate their basin treaties in order to avoid an increase in water insecurity (Goldenman 1990).

The final element is the concept of ambit, which describes and delimits the scope of water security – i.e., the sphere of influence of the notion. In addition to the traditional (hydrological and geographical) meaning of scope, the approach here is to better reflect the common challenges of water insecurity. So far, one of the main weaknesses of water cooperation is the inability to link various influencing factors in a comprehensive manner. The extent of the breadth of objectives covered by a freshwater agreement ranges from merely quantitative agreements to much more sophisticated institutions which also govern aspects of water quality and emergency situations. Evidently, the most effective management of transboundary watercourses, for the benefit of the whole basin, can only be achieved through a truly joint strategy involving all sectors and disciplines across borders (Magsig 2014). In addition to the predominant perception of scope, the element of ambit also does justice to the fact that water security has to be seen as a collective security issue (Magsig 2009). Owing to the interconnectedness of the globalized world and the role water plays in linking the various emerging crises, negative impacts may even be felt outside the basin. Thus, the times where water can solely be regarded as a national security issue are long gone, as one of our most fundamental common values is under threat – international peace and security. The linkages between different scales of cooperation over water (local, national, regional, and global) are fluid; and international law has to act as an interface between those layers while illustrating ways toward truly regional solutions.

5.2.2 The General Obligation to Cooperate

The question now arises how well international law accommodates this novel understanding of common water security. International environmental governance in general, and transboundary water management in particular, has long been dominated by the either/or debate on sovereignty versus the joint management of natural resources. While most states have now accepted a more nuanced interpretation of sovereignty, the debate about how sovereignty over freshwater resources should be interpreted today is still in full swing. Critically, the notion of sovereignty carries with it a responsibility to cooperate. As indicated by Article 1 of the UN Charter: "[t]he purposes of the United Nations are: […] (3) [t]o achieve international co-operation in solving international problems of an economic, social, cultural, or humanitarian character […]" (United Nations 1945). This unspecified duty to cooperate was partially clarified by the 1970 Declaration on Principles of International Law concerning Friendly Relations and Cooperation among States in accordance with the Charter of the United Nations, which stipulates that "states have the duty to co-operate with one another, irrespective of the differences in their political, economic and social systems, in the various spheres of international relations, in order to maintain international peace and security and to promote international economic stability and progress, the general welfare of nations and international co-operation free from discrimination based on such differences."

While the Declaration does not constitute binding international law, its universal recognition as a standard of conduct and perception of it as an elaboration of principles of international law give it considerable legal weight (Sands and Peel 2012). According to Judge Wolfrum, in the MOX case before the International Tribunal for the Law of the Sea, the duty to cooperate "balances the principle of sovereignty of states and thus ensures that community of interests are taken into account vis-à-vis individualistic state interests. It is the matter of prudence and caution as well in keeping with the overriding nature of the obligation to cooperate that the parties should engage therein [...]" (International Tribunal for the Law of the Sea 2001). It follows that there is a need to ultimately arrive at a stage where the concept of state sovereignty is understood as one of "cooperative sovereignty" (Perrez 2000). This necessity becomes particularly blatant when addressing the difficulty of managing common pool resources, where the collective action problem leads to unsatisfactory outcomes (Hardin 1968). Rather than treating sovereignty as a stumbling block in international negotiations – due to its apparent incompatibility with relinquishing freedoms and making concessions – acknowledging that the responsibility to cooperate is a key element of sovereignty itself seems to be a more promising strategy in addressing the tragedy of the commons (Schreuer 2002; Delbrück 2012). Hence, international law should provide a path for moving from "sovereignty as independence" to "sovereignty as interdependence."

In the arena of international water law, the general obligation to cooperate contains the procedural duties of prior information and of prior consultation, which aim to operationalize the rather vague principle. Yet, it still leaves a lot to be desired when it comes to fundamentally changing the way states perceive their national sovereignty over freshwater resources. One shortcoming of international water law is that states still have much discretion with regard to the particular means of cooperation. The setting up of joint institutions, for instance, is not compulsory, even though their immense benefit for transboundary freshwater management has long been proven (Schmeier 2013).

5.2.3 Regional Common Concern for Water Security

In order to strengthen the obligation to cooperate over shared freshwater resources, international law has to be developed further. The urgency to act jointly on more issues which bar unilateral action – like the management of transboundary freshwater resources – has led to the understanding of common security and revealed the limits of the current international legal regime. Earlier work has addressed the question as to how communality has been treated in international environmental law and what lessons can be learned for international water cooperation (Magsig 2014). While some approaches are too limited as they only apply to certain geographical areas beyond national jurisdictions and their resources (common area and common heritage), the notion of common concern appears to be the most promising in tackling issues of water security.

At the 1992 UN Conference on Environment and Development, a global framework for environmental responsibilities was designed which, for the first time, was based on a common concern, rather than the concept of good neighborliness (Birnie et al. 2009). This concern is based on the understanding that some kind of harm to the environment has the potential to adversely affect humanity as a whole; and thus, mitigating those impacts can only be achieved effectively if the international community in its entirety is involved. Acknowledging this position carries with it both a right and an obligation of the international community as a whole to have concern for the global environment (IUCN Commission on Environmental Law and International Council of Environmental Law 2010). Both treaties negotiated in 1992 in Rio – the UN Framework Convention on Climate Change (UNFCCC) and the Convention on Biological Diversity (CBD) – follow this approach. In its preamble, the CBD states that "the conservation of biological diversity is a common concern of humankind […]." While it lacks strong procedural support, the inclusion of common concern does ensure that the loss of biodiversity is being acknowledged as a major issue which the international community has to address on a common basis. However, this does not compensate for the shortcomings of the Convention – for instance, not extending state responsibility for extraterritorial harm to damage caused to the global commons (Guruswamy 1999). While the UNFCCC, which also features the concept of common concern in its preamble, kick-started a process of legal and political engagement to address global climate change, it too failed to overcome the state-centrism of the governance system, which ultimately led to watered-down obligations in weak agreements which were "designed to mask the political failure of the international community to create a global climate treaty" (Dimitrov 2010).

The Draft International Covenant on Environment and Development (2010) tried to redeem the concept's reputation, which was seemingly damaged by the two conventions' failing to implement it. Article 3 of the Covenant stipulates that "[t]he global environment is a common concern of humanity and under the protection of the principles of international law, the dictates of the public conscience and the fundamental values of humanity." According to the commentary, the concept should be interpreted as "the basis upon which the international community at all levels can and must take joint and separate action to protect the environment" (IUCN Commission on Environmental Law and International Council of Environmental Law 2010). In urging that not only single issues, like climate change or the loss of biodiversity, should be treated as being of common concern to the international community but also the environment as a whole, it constitutes a departure from previous approaches to common concern (Magsig 2015a). Given its lack of legitimacy, though, the Covenant has attracted only little attention among international legal scholars, despite it having breathed new life into the debate about communality in international law.

However, even in its embryonic stage, the concept of common concern is of particular interest to the advancement of international water law. Although its focus lies again on (common) benefits, it considers the benefits from common action rather than those derived from the mere exploitation of a resource (Brunnée 2007). Moreover, it fixes its attention on what renders a concern as being common, rather than targeting one particular area or resource, and thus avoids discussions about common property and territorial sovereignty (Magsig 2015a). One of its main advantages over other approaches to communality is that it triggers a shift from the orthodox reciprocity and material benefit sharing we often find in treaties of joint action in the long-term interest of the community (IUCN Commission on Environmental Law and International Council of Environmental Law 2010). Yet, the weaknesses of the notion are evident in the UNFCCC and the CBD, both facing difficulties in achieving strong legal impact with regard to the common concern at the level the conventions are intended to have an impact on – the global one. This is mainly due to the difficulties in phrasing a global set of values and interests which is detached from the individual interests of states. Even though, given the magnitude of the global water crisis, it should not be impossible to construct for transboundary freshwater cooperation an analogous mindset to the loss of biodiversity or climatic changes, the complexity of water security renders it impossible to agree on a perception of water security as a common concern of humankind (Magsig 2015a). Hence, scaling down one level by looking at the regional layer would be a fruitful middle road which can sufficiently accommodate the national interests and the global challenge. Here, the political and economic influence of regional institutions can be utilized in the push for change – change which is homegrown, rather than being perceived as imperialistic. Since, contrary to a river basin, a region is not a narrowly defined geographical area, framing water security as a regional common concern also opens up the enormous potential of including (non-state) actors and interests beyond the basin.

Thus, it may add a new dimension to international freshwater cooperation, which is still being perceived as a zero-sum game, by including non-riparian interests in the design and performance of international water cooperation (Brunnée and Toope 1994). On these grounds, the notion of a regional common concern is the most suitable conceptual vehicle for the endeavor to achieve water security (Magsig 2015a). While this does not necessarily require reinventing the wheel of transboundary freshwater cooperation, like the principle of equitable and reasonable utilization, it expects states to subscribe to certain basic rules and minimum managing standards as cornerstones for their transboundary water relations; and thus it triggers a process of rethinking focused on the respective regional approach. Hence, the concept of common concern provides a vehicle for inducting communality into international water law and arriving at more resilient agreements, since acknowledging that a particular challenge must be perceived as a matter of common concern results in the appreciation that transboundary water management can no longer be considered as a mere national issue. It shifts the responsibilities of states from individual to concerted action.

5.3 The Indus Waters Treaty and Its Application

The following section will now apply the developed water security lens to the Indus Waters Treaty and analyze how successful it is in strengthening transboundary freshwater cooperation and leading the way toward perceiving water security as being of regional common concern.

The conflict over the water resources of the Indus became international with the partition of British India, as the newly formed states were in disagreement over how to share and manage the previously unitary network of irrigation infrastructure (see also Chap. 1 of this book). The resultant power asymmetry between the two contracting parties has been identified as the main reason for the delayed completion of a water sharing agreement (Center for Policy and Human Development 2011). Here, the involvement of a third party, the World Bank, played a key role in continuously pushing negotiations forward (Biswas 1992). The fact that, after having been signed in 1960, the IWT survived three wars (1965, 1971, and 1999) between the two hostile neighbors (during which the water kept flowing) was reason enough for many scholars to celebrate the Treaty as a success (Khalid 2004). The question now arises whether the Indus Waters Treaty really satisfies a contemporary understanding of the duty to cooperate on transboundary waters and whether it contributes toward common water security in the Indus basin.

5.3.1 Availability

In terms of availability, the Indus Waters Treaty does not contain effective binding provisions addressing water quality or pollution. From the beginning of the negotiation process, the whole framework was focused on issues of quantity, apportioning the tributaries among the two nations. India's plan was to get all of the eastern rivers and 7 % of the western rivers, while Pakistan demanded 70 % of the eastern rivers and all of the western rivers (Biswas 1992). Ultimately, the parties agreed to allocate the tributaries with India receiving three tributaries – Sutlej, Beas, and Ravi (eastern rivers) – while Pakistan received the main Indus, Jhelum, and Chenab (western rivers) (Arts. II and III of the Indus Waters Treaty). However, the Indus Waters Treaty also allows India to tap the hydropower potential of the western rivers before they enter Pakistan (Art. III(2) of the Indus Waters Treaty) and guaranteed Pakistan a minimum quantity from the eastern rivers for a transitional period.

Yet, beyond the provisions of allocation, the IWT does not provide for meaningful obligations concerning availability. In lacking any effective mechanism dealing with environmental flows, ecosystem services, or demand management, the current legal framework does not provide for the sustainable control and protection of the Indus.

5.3.2 Access

The approach to resolving issues of access to water found in the Indus Waters Treaty is rather complex. Here, the principal institutional mechanism of the agreement, the Permanent Indus Commission, plays a significant role in the settlement of disputes, serving "as the regular channel of communication on all matters relating to the implementation of the Treaty" (Art. VIII of the IWT, "Permanent Indus Commission"). Under the agreement, issues that cannot be resolved by the Commission will be deemed "differences," which may, depending upon their classification, be heard by a "neutral expert" ("qualified engineer") at the request of either commissioner (Art. IX of the IWT). The difference will be considered to be a "dispute" if the matter falls outside those issues listed in Annex F. Disputes are to be resolved through negotiation and, failing any successful outcome, are subject to arbitration. This mechanism was triggered for the first time in the 45-year history of the treaty with a neutral expert asked to provide a determination of the difference regarding the Baglihar hydropower plant. In 2005, Pakistan contacted the World Bank stating that a "difference" had arisen with India under the Indus Waters Treaty relating to the Baglihar plant being constructed by India on the Chenab River in breach of the provisions under Paragraph 8 of Annex D to the Treaty. The bank appointed a neutral expert, who rendered a decision in February 2007. While some rightly argue that the differences were handled in a "transparent and fair manner" and acknowledge that the decision was accepted by the two parties (Salman 2008), the process did not manage to completely calm freshwater cooperation between India and Pakistan, nor did it give it a farsighted direction (Sinha 2010b). This becomes obvious when analyzing the outcome of the most recent legal dispute – the Kishanganga Arbitration (Rieu-Clarke et al. 2012; Uprety 2015). Following a Request for Arbitration by Pakistan in 2010, a dispute was (for the first time in the history of IWT) referred to a Court of Arbitration concerning India's construction of the Kishanganga Hydroelectric Project (KHEP). The design of the project is intended to divert waters from a damsite on the Kishanganga/Neelum River in the Jammu and Kashmir region to the Jhelum River – potentially reducing the power generation capacity of the planned Pakistani Neelum-Jhelum Hydroelectric Project (NJHEP). Pakistan identified two questions: (1) whether India's proposed diversion of the Kishanganga River for the run-of-the-river hydroelectric project into another tributary breaches India's legal obligations under Article III(2) of the IWT and (2) whether the agreement allowed India to deplete or bring the reservoir level of a run-of-river storage plant below the "Dead Storage Level" in circumstances other than unforeseen emergencies (Rieu-Clarke et al. 2012) (Fig. 5.1).

In September 2011, the Court issued an Order on Interim Measures, prohibiting India from constructing any permanent works on or above the Kishanganga/Neelum riverbed at the damsite that may inhibit the restoration of the full flow of the river to its natural channel (Permanent Court of Arbitration 2011). In February 2013, the Court issued a Partial Award, finding that India was permitted under the Treaty to divert water for the generation of electricity by the KHEP, arguing that Pakistan's water uses – of relevance here, the NJHEP – were preceded by the KHEP (Permanent Court of Arbitration 2013). Hence, the NJHEP was not considered an "existing use" under the Indus Waters Treaty which India was required to take into account at the time of planning its KHEP. The Court did, however, make clear that India's right to divert the waters of the Kishanganga/Neelum was not absolute, since relevant principles of customary international law, including principles of international environmental law, have to be taken into account. Accordingly, the award allows India to proceed with the construction of the KHEP, subject to ensuring a minimum downstream flow to be determined in the Final Award. Further, it prohibits India from using drawdown flushing for sediment control at the Kishanganga Project and any future run-of-river plant on the western rivers. On 20 December 2013, the Permanent

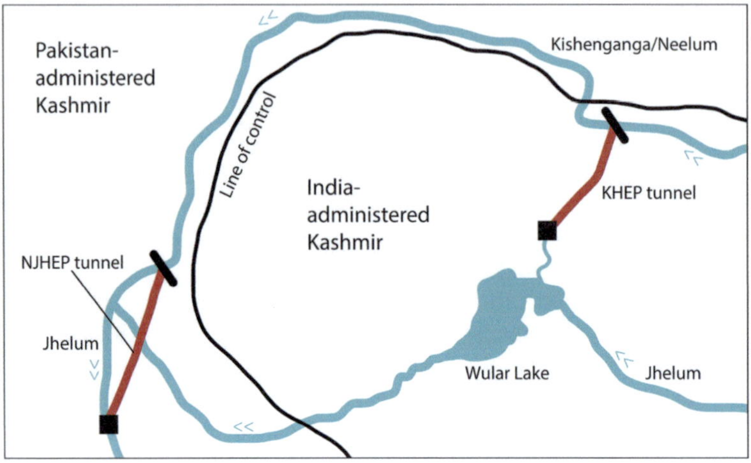

Source: (Magsig 2015a, p. 187)

Fig. 5.1 The Kishanganga arbitration (Source: Magsig 2015a, p. 187)

Court of Arbitration rendered its Final Award (Permanent Court of Arbitration 2013), stating that in deciding the rate of minimum flow, it was necessary to "mitigate adverse effects to Pakistan's agricultural and hydro-electric uses throughout the operation of the KHEP, while preserving India's right to operate the KHEP and maintaining the priority it acquired from having crystallized prior to the NJHEP" and to give due regard to "the customary international law requirements of avoiding or mitigating trans-boundary harm and of reconciling economic development with the protection of the environment."

While the decision "serves as a useful reminder of the potential of pacific dispute settlement in resolving complex disputes in tense settings" (Kumar 2013), it, at least to some extent, contradicts the outcome of the Baglihar difference. While the Baglihar judgment allowed India to draw down water below the dead storage level under certain conditions and apply a technique called drawdown flushing in order to protect the hydropower plant from siltation, the Permanent Court of Arbitration, of course, did not treat the Baglihar findings and outcome as a precedent. To the contrary, it decided that from now on this technique shall be prohibited. Interestingly, the Court's use of customary international law in the Kishanganga Arbitration was limited by Paragraph 29 of Annex G to the Indus Waters Treaty. While in its Partial Award it emphasized that it was "incumbent upon [it] to interpret and apply this 1960 Treaty in light of the customary international principles for the protection of the environment in force today," in the Final Award the Court qualified this duty by arguing that "if customary international law were applied not to circumscribe, but to negate rights expressly granted in the Treaty, this would no longer be 'interpretation or application' of the Treaty but the substitution of customary law in place of the Treaty" (Permanent Court of Arbitration 2013). Thus, it seems illusive to expect an end of disputes revolving around the same issues, as – different to a decision by a court which can set a precedent – the next ruling of a Court of Arbitration will depend on many unforeseeable factors – including the composition of the arbitral tribunal.

5.3.3 *Adaptability*

Due to climatic changes and pressures of population growth, the Indus basin faces huge challenges in terms of the adaptability of the regime (see Chap. 2 of this book). The whole Treaty resembles more of a divorce settlement, rather than a future-proof agreement for the sustainable management of a river basin. However, it is important to acknowledge the fact that the regime does not exist in isolation – and that international law does not exist without interpretation (Hollis 2014). As the Kishanganga Arbitration has shown, new conflicts about the utilization of the shared water resources of the Indus will have to take recent developments of international law into consideration – at least to an extent which is considered within the means of treaty interpretation. The fact that there is no consensus on the methodology of treaty interpretation, however, might cause even more uncertainty among the parties (Bjorge 2014; Villiger 2011).

Since the Indus Waters Treaty fails to accommodate flexible mechanisms, other than the general rules of interpretation, it has been argued that the Treaty should be amended to better cope with climate change uncertainty (Bagla 2010).

5.3.4 *Ambit*

While the Indus river basin is shared between Pakistan (47 %), India (39 %), China (8 %), and Afghanistan (6 %), the Treaty does not involve the latter two nations (AQUASTAT 2011). This major shortcoming does not prevent the majority of scholars to still view the Indus Waters Treaty as a success – like McKinney (2011) calling it "one of the most successful settlements of a transboundary water basin conflict." Yet, when looking at the actual ambit of the agreement, one has to follow the minority view that the Treaty can only be perceived as a disappointment. The reason for the dramatically diverging views on the quality of the agreement might originate from asking a completely different question. While most commentators seem to be satisfied with the "survival" of the Treaty during times of war, should one not be able to demand more from a treaty governing a shared watercourse – a vital resource for both countries? Is it too bold to ask why the Indus Waters Treaty did not prevent three wars? In claiming that "[t]he Indus Waters Treaty, which is the most successful India-Pakistan agreement to date, has held up for 46 years largely because the Treaty does not require daily interaction and joint decision making by those two estranged governments" (Schaffer 2007), do we imply that we consider treaties which do not require adequate cooperation to be a success? International legal scholarship would certainly damage itself if it followed such an absurd understanding of international law.

5.4 The Way Forward

Following a contemporary understanding of common water security, based on the notion of hydrosolidarity, a freshwater treaty should not merely be able to muddle through difficult bilateral times. It should provide an impetus for the riparian countries to develop relations outside the water box – and ultimately lead to more peaceful relations. In this regard, as well as concerning the vital issues of water quality and adaptability, the Indus Waters Treaty has missed an important opportunity. While it has been argued before that renegotiations are inevitable (Bhatnagar 2009; Sinha 2010a), the examination through the contemporary security lens has made it even more obvious that the agreement between India and Pakistan in its current form has no future, as it does not address water security as a regional common concern. The question now arises as to how the normative pillars should be modernized in order to build a more resilient future for the Indus basin. Following a path within the unmodified regime where the contemporary interpretation of international water law will be applied to the rigid form of the Indus Waters Treaty cannot be regarded a future-proof option.

5.4.1 Options Within the Existing Framework

When criticizing the Indus Waters Treaty for its lack of flexibility and inability to strengthen cooperation between India and Pakistan, one has to acknowledge that the treaty text does provide some room for maneuver. This room, however, has not yet been utilized by either party. Article VII(1) of the IWT states that "[t]he two Parties recognize that they have a common interest in the optimum development of the Rivers, and, to that end, they declare their intention to co-operate, by mutual agreement, to the fullest possible extent."

In theory, this provision opens the door for various tools of cooperation in addition to the Indus Waters Treaty and without having to touch the hot potato of renegotiations. One could, for example, imagine both states to see the urgent need to address very specific challenges where states might already see the benefit in joint and coordinated action. Here, the development of common obligations regarding emergency response mechanisms immediately comes to mind. This could easily happen within the existing legal framework based on Article VII of the IWT and would immediately benefit millions of people who are regularly hit by water-related disasters in the basin. While it is disappointing to note that no initiatives have been undertaken under the provision of "future cooperation," it does not really come as a great surprise. Given that it carries forward the logic of partition of the two countries to their shared water resources, the IWT does not treat the Indus as a single unit – not promoting any cooperation which goes beyond what was agreed upon in 1960.

5.4.2 *Renegotiating the Indus Waters Treaty*

Given the fact that India and Pakistan have failed to exploit the provisions of the IWT which are supposed to strengthen international cooperation (Art. VII), the current framework does not seem to trigger the needed change in political will. Hence, it seems obvious that the best option for achieving water security in the Indus basin is amending some of the terms of the Indus Waters Treaty or renegotiating an entirely new agreement. For such a process, the analysis in Chap. 1 can be used as a thought-provoking impulse. To begin with, a renegotiated agreement should be brokered between all basin states – including Afghanistan and China. Further, covering issues of water quality would most certainly improve the situation on the ground. Here, the concept of environmental flows should be incorporated (Forslund et al. 2009). Furthermore, the wider benefits from the utilization of the shared water resources have to be put onto the negotiation table as well. Benefit sharing, in particular concerning the need for food and energy security in the basin, promises an increase in acceptance of the legal rules and the building of trust among co-riparians (Wouters and Moynihan 2013; Ziganshina 2014). In order to be more resilient, any new regime governing the Indus has to be reasonably flexible and thus allow for dealing with the various uncertainties in transboundary water cooperation (McCaffrey 2003). Here, setting up a joint river basin organization which is in a position to not only make recommendations to member states, but actually decide swiftly – and independently from the political quarrels – on matters of transboundary water management, would certainly inject a huge amount of flexibility into the legal framework (Schmeier 2013).

For the arduous task of renegotiating the IWT, the UN Watercourses Convention (United Nations 1997) could serve as a valuable starting point, as its primary purpose as a global framework instrument is to supplement existing regional (multi-basin), basin, and subbasin agreements. In particular, the Convention can assist in filling gaps where existing water agreements fall short, as it includes several rules of customary international law – e.g., equitable and reasonable utilization – and gives guidance as to how to implement them (Rieu-Clarke et al. 2012). However, the perception of the UN Watercourses Convention in Himalayan Asia does reveal certain difficulties for using the Convention as a potential blueprint for basin agreements. One of this region's countries, China, voted against the adoption of the Convention, five of them (Afghanistan, Bhutan, Burma, India, and Pakistan) abstained or were absent, and six of them (Bangladesh, Cambodia, Laos, Nepal, Thailand, and Vietnam) voted in favor, with only one of the latter, Vietnam, having ratified the Convention until now. More work is needed in order to address the misconceptions of many Asian countries regarding the UN Watercourses Convention (Loures and Rieu-Clarke 2013).

In any case, the spirit of the IWT agreement has to be changed from a divorce contract to a joint vision for common water security. Rather than the shortsighted "react-and-correct" approach, one of "foresee and prevent" is desperately needed (Timoshenko 1992). In order to be successful, the renegotiated agreement would have to be considered a new starting point of water cooperation in the basin, rather than the resolution of a conflict.

5.4.3 The Chances of a Regional Framework

The contemporary understanding of water security has also brought about renewed interest in regionalism in international water law (Moynihan and Magsig 2014). How can regional approaches to transboundary water cooperation be developed in order to be conducive to a broader regional governance framework aiming at the development of concrete forms of integrated cooperation on a whole range of transboundary environmental issues?

At the same time as most countries in Himalayan Asia have seen their renewable freshwater resources and water availability drop continuously over the last decades (Asian Development Bank 2013), regional cooperation is being hampered by political tensions between several states – e.g., the conflicts in the volatile parts of Kashmir and Tibet. The immense pressures of decreasing water quality and increasing competition for freshwater not only affect the states in their respective national development but also transform these domestic challenges into regional ones. The glaciers of the Himalayas feed the headwaters of the mighty rivers Yellow, Yangtze, Mekong, Salween, Irrawaddy, Ganges-Brahmaputra, and Indus, in which more than 1.5 billion people directly depend on (Grey and Connors 2009). China's and India's emphases on large-scale infrastructure in addressing their water issues – the South-North Water Diversion Project (Berkoff 2003) and the River Linking Project (Khalid 2004), respectively – have huge implications for their downstream neighbors. The governments' ambitious plans to step up hydropower capacity and push forward with interbasin water transfers will certainly increase the geopolitical risks of international freshwater cooperation in Himalayan Asia.

When looking at China, treaty practice does not allow for excessive optimism, either, as all freshwater agreements China has entered are bilateral, despite the fact that many of them govern multistate watercourses (Wouters and Chen 2013). While there are some success stories of international water law in Himalayan Asia – e.g., (at least to some extent) in the Mekong basin (Rieu-Clarke and Gooch 2010) – the region remains rather hostile toward the idea of a more common approach to water security. Yet, a gradual development toward closer regional cooperation on freshwater issues is by no means illusive (Magsig 2015b). Here, regional organizations – e.g., the South Asian Association for Regional Cooperation (SAARC) and the Association of Southeast Asian Nations (ASEAN) – can play an important role, as they allow for gradual strengthening and deepening of relations on a diverse range of topics, including water cooperation, not only in the Indus basin but the whole of

Himalayan Asia (see also Chap. 8 of this book). It is in both regional hegemons' interest to work toward a resilient future of the region, as this will allow both China and India to sustain their economic growth. In order to do so, however, they have to arrive at a common understanding of their water insecurities and thus employ compatible – if not common – strategies in addressing them (Tellis and Mirski 2013). This need clearly makes the relationship between China and India the decisive factor in the future water security of the region (see Chap. 10 of this book).

Himalayan Asia will not eagerly implement a regional approach to water cooperation which comes from outside the region, since third-party involvement – e.g., by the World Bank – is being seen rather critically. It is much more likely that states within Himalayan Asia will develop their own concepts – based on their regional identity and specific political and cultural environment. However, in developing novel pathways toward an understanding of cooperative sovereignty, lessons from other regional regimes can be most helpful (Moynihan and Magsig 2014).

5.4.4 *Observations*

A contemporary understanding of the duty to cooperate over transboundary waters implies more than merely dividing the tributaries of an international watercourse like the Indus. International law is gradually moving toward a regime which evokes shared responsibilities – and, thus, is able to address common concerns like water insecurity more effectively. By reassessing the Indus Waters Treaty against this backdrop, its weaknesses have been revealed. Here, the concept of considering water security as a matter of regional common concern was introduced as a promising way forward in striving for truly joint and long-term regional water management. If taken seriously, this may even involve a process which goes beyond the basin. While there is justified skepticism about the role regionalism can play in effectively addressing water cooperation in Himalayan Asia, the urgency of the crisis could soon enable the environment for a regional approach which addresses water security from a common concern perspective – and leads to shared responsibilities (Moynihan and Magsig 2014).

In the meantime, important first steps toward a more resilient future of the Indus basin can and should be pursued at a bilateral level – both within the existing legal framework and by amending the IWT. Those two strategies do by no means have to happen in isolation. To the contrary, India and Pakistan are well advised to take immediate steps under Article VII of the IWT to address very specific challenges where states already see immediate benefits from joint action – like the development of common obligations regarding emergency response mechanism. Simultaneously, they should engage in more long-term water diplomacy activities to start discussing the future of their legal regime.

As has been demonstrated by Swain in Chap. 3 of this book, the costs of noncooperation in the basin are immense. The time to act is now. The longer the basin states shy away from addressing the underlying issues of their legal framework, the more difficult it will get to remedy its shortcomings. Eventually, India and Pakistan will both see the need to put their legal agreement on the Indus on a new path which will allow them to exploit the enormous potential of sharing the benefits of cooperation. Until then, however, it seems unlikely that the two countries will agree to modify the IWT and turn it from a water portioning agreement into a contemporary water resource development one (Sinha et al. 2012).

It is hoped that the initiated discussion about perceiving water security as a regional common concern triggers fundamental change in how states cooperate on the highly complex and controversial issues concerning their shared freshwater resources in the future. International law can provide the framework for imagining an "Industan" which moves beyond the prevalent state-centric approaches to international cooperation over freshwater resources.

References

Abseno MM (2013) The influence of the UN Watercourses Convention on the development of a treaty regime in the Nile river basin. Water Int 38(2):192–203. doi:10.1080/02508060.2013.782798

Allan T (2002) The Middle East water question: hydropolitics and the global economy. I.B. Tauris, London/New York

AQUASTAT of FAO (2011) Indus river basin. http://www.fao.org/nr/water/aquastat/basins/indus/index.stm

Asian Development Bank (2013) Asian water development outlook 2013: measuring water security in Asia and the Pacific. Asian Development Bank, Manila

Bagla P (2010) Along the Indus river, saber rattling over water security. Science 328(5983):1226–1227

Barnaby W (2009) Do nations go to war over water? Nature 458(7236):282–283

Berkoff J (2003) China: the South-North water transfer project: is it justified? Water Policy 5(1):1–28

Bhatnagar M (2009) Reconsidering the Indus waters treaty. Tulane Environ Law J 22(2):271–314

Birnie P, Boyle A, Redgwell C (2009) International law and the environment, 3rd edn. Oxford University Press, Oxford

Biswas AK (1992) Indus Water Treaty: the negotiating process. Water Int 17(4):201–209

Bjorge E (2014) The evolutionary interpretation of treaties. Oxford University Press, Oxford

Brennan JF (2008) The China-India-Pakistan water crisis: prospects for interstate conflict. Naval Postgraduate School, Monterey

Brock H (2011) Competition over resources: drivers of insecurity and the Global South. Oxford Research Group, London

Brunnée J (2007) Common areas, common heritage, and common concern. In: Bodansky D, Brunnée J, Hey E (eds) The Oxford handbook of international environmental law. Oxford University Press, Oxford, pp 550–573

Brunnée J, Toope SJ (1994) Environmental security and freshwater resources: a case for international ecosystem law. Yearb Int Environ Law 5:41–76

Buzan B (1991) People, states and fear: an agenda for international security studies in the post-cold war era, 2nd edn. Harvester Wheatsheaf, London

Center for Policy and Human Development (2011) Afghanistan human development report 2011. The forgotten front: water security and the crisis in the sanitation. Center for Policy and Human Development, Kabul

Chellaney B (2011) Water: Asia's new battleground. Georgetown University Press, Washington, DC

Delbrück J (2012) The international obligation to cooperate: an empty shell or a hard law principle of international law? A critical look at a much debated paradigm of modern international law. In: Hestermeyer HP, König D, Matz-Lück N, Röben V, Seibert-Fohr A, Stoll T-P et al (eds) Coexistence, cooperation and solidarity, vol 1. Martinus Nijhoff Publishers, Leiden, pp 3–16

Dimitrov RS (2010) Inside UN climate change negotiations: the Copenhagen conference. Rev Pol Res 27(6):795–821. doi:10.1111/j.1541-1338.2010.00472.x

Expert determination on points of difference referred by the Government of Pakistan under the provisions of the Indus Water Treaty, Executive Summary (12 February 2007). Available at http://siteresources.worldbank.org/SOUTHASIAEXT/Resources/223546–1171996340255/BagliharSummary.pdf

Forslund A, Malm-Renöfält B, Barchiesi S, Cross K, Davidson S, Farrell T et al. (2009) Securing water for ecosystems and human well-being: the importance of environmental flows. Swedish Water House Report, 24

Fröhlich CJ (2012) Security and discourse: the Israeli-Palestinian water conflict. Confl Secur Dev 12(2):123–148. doi:10.1080/14678802.2012.688290

Gleditsch NP (1998) Armed conflict and the environment: a critique of the literature. J Peace Res 35(3):381–400

Goldenman G (1990) Adapting to climate change: a study of international rivers and their legal arrangements. Ecol Law Q 17(4):741–802

Grey D, Connors G (2009) The water security imperative: we must and can do more. 5th World Water Forum Istanbul 2009. World Water Council, Istanbul, pp 58–62

Guruswamy LD (1999) The convention on biological diversity: exposing the flawed foundations. Environ Conserv 26(2):79–82

Hardin G (1968) The tragedy of the commons. Science 162(3859):1243–1248

Hollis DB (2014) The existential function of interpretation in international law. In: Bianchi A, Peat D, Windsor M (eds) Interpretation in international law. Oxford University Press, Oxford, pp 78–110

Homer-Dixon TF (1994) Environmental scarcities and violent conflict: evidence from cases. Int Secur 19(1):5–40

ICJ (1997) Case Concerning the Gabčíkovo-Nagymaros Project (Hungary v. Slovakia) General List no. 92. Reprinted in 37 ILM 162 (1998)

International Tribunal for the Law of the Sea (2001) The MOX plant case (Ireland v. United Kingdom). Found at: <http://www.itlos.org/fileadmin/itlos/documents/cases/case_no_10/Order.03.12.01.E.pdf>

IUCN Commission on Environmental Law, & International Council of Environmental Law (2010) Draft international covenant on environment and development, 4th edn. IUCN, Gland

Khalid ARM (2004) The interlinking of rivers project in India and international water law: an overview. Chin J Int Law 3(2):553–570

Kumar S (2013) The Indus waters Kishenganga arbitration (Pakistan v. India). ASIL Insights 17(13):1–8

Loures FR, Rieu-Clarke A (eds) (2013) The UN watercourses convention in force: strengthening international law for transboundary water management. Earthscan, London

Magsig B-O (2009) Introducing an analytical framework for water security: a platform for the refinement of international water law. J Water Law 20(2/3):61–69

Magsig B-O (2014) Pushing the boundaries: rethinking international law in light of the common concern for water security. In: Sancin V, Kovič Dine M (eds) International environmental law: contemporary concerns and challenges in 2014. GZ Založba, Ljubljana, pp 441–452

Magsig B-O (2015a) International water law and the quest for common security. Routledge, London

Magsig B-O (2015b) Water security in Himalayan Asia: first stirrings of regional cooperation? Water Int 40(2):342–353

Mandhanda N for Time Magazine (2012) Water wars: why India and Pakistan are squaring off over their rivers. http://www.time.com/time/world/article/0,8599,2111601,00.html. 16 April 2012
McCaffrey SC (2003) The need for flexibility in freshwater treaty regimes. Nat Res Forum 27(2):156–162
McKinney DC (2011) Transboundary water challenges: case studies. Center for Research in Water Resources, University of Texas, Austin
Moynihan R, Magsig B-O (2014) The rising role of regional approaches in international water law: lessons from the UNECE water regime and Himalayan Asia for strengthening transboundary water cooperation. Rev Eur Commun Int Environ Law 23(1):43–58
Orakhelashvili A (2011) Collective security. Oxford University Press, Oxford
Permanent Court of Arbitration (2013) Final award in the matter of the Indus waters Kishenganga arbitration (Pakistan v. India). http://www.pca-cpa.org/showfile.asp?fil_id=2471. 20 December 2013
Permanent Court of Arbitration (2011) Order on interim measures in the matter of the Indus waters Kishenganga arbitration (Pakistan v. India). http://www.pca-cpa.org/showfile.asp?fil_id=1726. 23 September 2011
Permanent Court of Arbitration (2013). Partial award in the matter of the Indus waters Kishenganga arbitration (Pakistan v. India). http://www.pca-cpa.org/showfile.asp?fil_id=2101. 18 February 2013
Perrez FX (2000) Cooperative sovereignty: from independence to interdependence in the structure of international environmental law. Kluwer Law International, The Hague/Boston
Rieu-Clarke A, Gooch G (2010) Governing the tributaries of the Mekong: the contribution of international law and institutions to enhancing equitable cooperation over the Sesan. Pac McGeorge Glob Bus Dev Law J 22(2):193–224
Rieu-Clarke A, Moynihan R, Magsig B-O (2012) UN watercourses convention: user's guide. IHP-HELP Centre for Water Law, Policy and Science, Dundee
Salman SMA (2006) International water disputes: a new breed of claims, claimants, and settlement institutions. Water Int 31(1):2–11
Salman SMA (2008) The Baglihar difference and its resolution process: a triumph for the Indus Waters Treaty? Water Policy 10(2):105–117
Sands P, Peel J (2012) Principles of international environmental law, 3rd edn. Cambridge University Press, Cambridge
Schaffer TC (2007) Putting the Kashmiris into the Kashmir issue. Asia Pol 3:192–195
Schmeier S (2013) Governing international watercourses: river basin organizations and the sustainable governance of internationally shared rivers and lakes. Routledge, New York
Scholtz W (2009) Collective (environmental) security: the yeast for the refinement of international law. Yearb Int Environ Law 19:135–162
Schreuer C (2002) State sovereignty and the duty to cooperate – two incompatible notions? In: Delbrück J (ed) International law of cooperation and state sovereignty: proceedings of an international symposium of the Kiel Walther Schücking-Institute of International Law, May 23–26, 2001. Duncker & Humblot, Berlin, pp 163–180
Schuster-Wallace CJ, Grover VI, Adeel Z, Confalonieri U, Elliott S (2008) Safe water as the key to global health. United Nations University, International Network on Water, Environment and Health, Hamilton
Sinha UK (2010a) 50 years of the Indus Water Treaty: an evaluation. Strateg Anal 34(5):667–670
Sinha UK (2010b) Water a pre-eminent political issue between India and Pakistan. Strateg Anal 34(4):482–485
Sinha UK, Gupta A, Behuria A (2012) Will the Indus Water Treaty survive? Strateg Anal 36(5):735–752. doi:10.1080/09700161.2012.712376
Taenzler D, Ruettinger L, Ziegenhagen K, Murthy G (2011) Water, crisis and climate change in India: a policy brief. Adelphi, Berlin
Tellis AJ, Mirski S (eds) (2013) Crux of Asia: China, India, and the emerging global order. Carnegie Endowment, Washington, DC

Timoshenko AS (1992) Ecological security: response to global challenges. In: Weiss EB (ed) Environmental change and international law: new challenges and dimensions. United Nations University Press, Tokyo, pp 413–458

Treaty between India and Pakistan Regarding the Use of the Waters of the Indus (19 September 1960; entered into force 1 April 1960) 419 UNTS 125 (1960)

UN General Assembly, Declaration of Principles of International Law Concerning Friendly Relations and Co-operation Among States in Accordance with the Charter of the United Nations (UNGA Resolution A/RES/2625(XXV), 24 October 1970)

United Nations (1945) Charter of the United Nations (San Francisco, 26 June 1945; in force 24 October 1945)

United Nations (1997) Convention on the Law of the non-navigational uses of international watercourses (New York, 21 May 1997; in force 17 August 2014)

United Nations (2004) A more secure world: our shared responsibility – report of the secretary-general's high-level panel on threats, challenges and change. United Nations, New York

Uprety K (2015) The Kishenganga arbitration: reviving the Indus Treaty and managing transboundary hydropolitics. [Article]. Chin J Int Law 14(3):497–543. doi:10.1093/chinesejil/jmv029

Villiger ME (2011) The rules on interpretation: misgivings, misunderstandings, miscarriage? The 'crucible' intended by the International Law Commission. In: Cannizzaro E (ed) The law of treaties beyond the Vienna Convention. Oxford University Press, Oxford, pp 105–122

Voigt C (2009) Sustainable security. Yearb Int Environ Law 19:163–196

von Tigerstrom B (2007) Human security and international law. Hart Publishing, Oxford

Weinthal E, Zawahri N, Sowers J (2015) Securitizing water, climate, and migration in Israel, Jordan, and Syria. Int Environ Agreements: Polit Law Econ 15(3):293–307. doi:10.1007/s10784-015-9279-4

Wirsing RG, Stoll DC, Jasparro C (2013) International conflict over water resources in Himalayan Asia. Palgrave Macmillan, Basingstoke

Wolpert S (2010) India and Pakistan: continued conflict or cooperation? University of California Press, Berkeley

Wouters P, Chen H (2013) China's 'soft-path' to transboundary water cooperation examined in the light of two UN global water conventions – exploring the 'Chinese way'. J Water Law 22(2/3):229–247

Wouters P, Moynihan R (2013) Benefit sharing in the UN Watercourses Convention and under international water law. In: Loures FR, Rieu-Clarke A (eds) The UN Watercourses Convention in force: strengthening international law for transboundary water management. Earthscan, London, pp 321–335

Wouters P, Vinogradov S, Allan A, Jones P, Rieu-Clarke A (2005) Sharing transboundary waters: an integrated assessment of equitable entitlement: the legal assessment model, vol 74, IHP-VI technical documents in hydrology. UNESCO, Paris

Wouters P, Vinogradov S, Magsig B-O (2009) Water security, hydrosolidarity and international law: a river runs through it …. Yearb Int Environ Law 19(1):97–134

Zhang Y, Block P, Hammond M, King A (2015) Ethiopia's Grand Renaissance Dam: implications for downstream riparian countries. J Water Resour Plan Manag 141(9):05015002_05015001–05015002_05015010, doi:10.1061/(ASCE)WR.1943-5452.0000520

Ziganshina D (2014) Promoting transboundary water security in the Aral Sea basin through international law (International Water Law). Brill Nijhoff, Leiden

Chapter 6
Managing the Indus in a Warming World: The Potential for Transboundary Cooperation in Coping with Climate Change

David Michel

Abstract Decision-makers in Afghanistan, China, India, and Pakistan must reconcile a host of overlapping socioeconomic, ecological, and policy pressures to ensure their countries' future water needs. Growing populations and expanding economies are driving rising water demands, even as environmental degradation and unsustainable consumption practices increasingly stretch the shared resources of the Indus River. Climate change will compound the challenges confronting water managers. None of the riparian states can successfully surmount these tests on its own. Greater dialogue and coordination among the basin's diverse communities offers considerable scope for mitigating mutual threats and generating collective benefits. Collaborative approaches promoting data exchange, capacity building, and knowledge generation can help policymakers and the broader public better apprehend and assess the basin's complex climate and water challenges. Common frameworks for identifying and adopting policy lessons can enlarge the range of policy choices, scale up best practices, and chart cooperative pathways forward.

Keywords Indus • Climate change • Cooperation • Integrated water resources management

6.1 Introduction

Global climate change will significantly impact freshwater resources across the Indus Basin.

Continuing global warming threatens to upset the prevailing regional precipitation patterns, shuffling the seasonal timing and geographical distribution of the rain and

D. Michel (✉)
The Stimson Center, Washington, DC, USA

The Geneva Centre for Security Policy, Geneva, Switzerland
e-mail: dmichel@stimson.org

Z. Adeel, R.G. Wirsing (eds.), *Imagining Industan*, Water Security in a New World, DOI 10.1007/978-3-319-32845-4_6

snowfalls that sustain the basin's water supplies. Rising global temperatures are gradually shrinking the Himalayan glaciers whose summer meltwaters feed the Indus, foreshadowing future shortfalls in their contribution to the river's flow. Complex computer analyses calibrated to model growing stresses on the Earth's climate systems project that extreme weather events will increase in frequency and degree around the basin, with stronger storms, higher floods, and deeper droughts becoming more numerous and severe. Such accumulating impacts, combining both acute disasters and chronic stresses, risk imposing substantial strains on the vital water ecosystems, infrastructure, and services on which the Indus nations critically rely.

Decision-makers in Afghanistan, China, India, and Pakistan must reconcile a host of overlapping socioeconomic, ecological, and policy pressures as they endeavor to ensure their countries' future water needs and safeguard the resources of the Indus River they all share. Growing populations, rapid urbanization, and expanding economies are driving rising water demands, even as environmental degradation and unsustainable consumption practices increasingly stretch available water supplies. Climate change will compound the challenges confronting water managers. None of the riparian states can successfully surmount these tests entirely on its own. Global warming heeds no borders. Its effects will be felt throughout the basin. The Indus' hydrology similarly ignores national boundaries. Like the tributaries of the river itself, the water resources and the water policies of the riparian states ultimately intertwine. Management choices made by one country can either enhance water benefits or undermine water security for its neighbors. To meet the collective dangers of climate change, the peoples of "Industan" must imagine and enact more collaborative water governance strategies.

6.2 Growing Climate Pressures on Indus Basin Water Resources

Global climate change threatens to disrupt freshwater systems worldwide. The Earth's climate and its hydrological processes are closely interconnected in complex ways. Both are sensitive to perturbations by human influences. Global warming will accelerate the hydrologic cycle. As temperatures rise, so does the ability of the atmosphere to hold water, increasing precipitation and evaporation and impinging on fundamental hydrometeorological mechanisms. Elemental patterns such as the onset of the monsoon and the recurrence of El Niño-Southern Oscillation phenomena may shift or falter. Globally, the Intergovernmental Panel on Climate Change (IPCC) projects mounting flood and drought risks over the twenty-first century. Long-term variations in the volume, intensity, timing, location, and form of precipitation (whether it falls as rain or as snow) could substantially affect the freshwater supplies available to communities and ecosystems (Jiménez Cisneros et al. 2014). According to one analysis, by 2050, climate change will alter river flows in every populated basin on the planet (Palmer et al. 2008).

Global warming is already impacting the Earth's climate. In its extensive Fifth Assessment Report on the science of climate change, the IPCC determined that global average surface temperatures have warmed 0.85 °C over the period 1880–2012. Each of the past three decades has been successively hotter than any previous decade in the instrumental record, with the 2000s being the warmest decade of all. The available evidence suggests global average annual precipitation has also increased, rising by a few millimeters per decade from 1901 to 2008. Extreme weather events occur more frequently. Since the mid-twentieth century, heat waves strike more often and are lasting longer around much of the world. Similarly, intense precipitation events – heavy rainfalls or snowfalls that would normally occur only 5 % of the time – are happening more often (Hartmann et al. 2013).

6.2.1 Observed Climate Trends

Underneath the global averages, the regional impacts of climate change vary widely. Meteorological records track temperatures rising 1–1.5 °C around the Indus from 1901 to 2012, the most significant warming appearing in the mountainous upper basin. Precipitation data display less clear tendencies, but broadly manifest an uptick in annual and seasonal precipitation in recent decades (Hartmann et al. 2013; Hijioka et al. 2014; Nepal and Shrestha 2015). Extreme high temperatures have been recurring somewhat more often since 1951. Extreme precipitation, meanwhile, exhibits considerable contrasts. During 1979–2011, the frequency of extreme events dropped notably in the western basin (including Afghanistan) but climbed markedly in the north and east (including China) (Hartmann and Buchanan 2014).

National data furnish finer geographical resolution on observed trends, illuminating complex spatial and temporal patterns. In Afghanistan, where the Kabul River forms a major tributary of the Indus, decades of instability have disrupted climate and hydrological data collection. Nevertheless, studies by the United States Geological Survey show average winter temperatures (February) in the Kabul Basin rising at a rate of 2 °C per decade since the early 1960s. Sparse stream gauge data suggests the mounting temperatures may be affecting river flows, displacing the period of peak runoff into the spring as snow and ice fields melt earlier in the warmer climate (Mack et al. 2010). For western China, where the Indus main stem and the Sutlej tributary rise, what meager information exists points to slight temperature increases of 0.01–0.04 °C per year and a similarly slender drop in precipitation (Singh et al. 2011). In India, both mean annual temperatures and maximum temperature trends in the northern Indus Basin states of Himachal Pradesh and Jammu and Kashmir have risen over 1901–2007, with areas of accelerating warming since 1971 expanding to encompass Punjab as well. Jammu and Kashmir has experienced statistically significant increases in extreme rains in the eastern half, but falling trends in the west (Ministry of Environment and Forests 2010, 2012). In Pakistan, annual average temperatures warmed in the upper Indus and Punjab plains over the past 50 years but cooled in the lower Indus. Springs (April to May) have been hotter through most of the basin, but the monsoons cooler. Annual average precipitation has jumped 25 %, or 63 mm, during the twentieth century. The upper Indus, central Punjab, and Northern Balochistan witnessed the greatest increase throughout the year, while the high Himalaya and the lower Indus saw modest declines in winter precipitation (Yu et al. 2013).

Climate change will exert additional, chronic pressures on key determinants of freshwater supplies in the Indus. Some 50 % of the precipitation nourishing the region falls during the summer (June to September) monsoon, often in sudden downpours (Rajbhandari et al. 2015). The monsoon's strength and timing depend on the moisture content of the atmosphere, the temperature contrast between the ocean and neighboring land surface, and related factors such as continental snow and ice cover. Climate change will impact all these drivers, raising concerns that global warming could scramble the monsoon regime. Reviewing the twentieth-century record, the IPCC concluded no systematic long-term trends have yet appeared in the behavior of the South Asian monsoon (Christensen et al. 2013). Within the Indus Basin, however, a more variegated picture emerges. Recent analyses of India's 1901–2012 monsoon seasons revealed precipitation increasing in an arc stretching from Punjab through northern Jammu and Kashmir, but falling in Himachal Pradesh (Roxy et al. 2015). Another study delving even deeper into the meteorological archives, from 1871 to 2005, showed northwestern India increasingly subject to both extreme deficit and extreme excess monsoon rainfalls (Pal and Al-Tabaa 2010). In Pakistan, work by the Global Change Impact Study Centre and the Pakistan Meteorological Department found rising monsoon season precipitation trends since 1951 over most of the basin except the coastal zone, with extreme precipitation events increasing across the county (Sheikh et al. 2009; Imran et al. 2014).

More than any other major river system, the Indus depends on snow and ice melt to sustain its flow. The Indus headwaters rise in the mountain glaciers of the Hindu Kush Himalaya (HKH). Often called the continent's "water towers," the Himalayan glaciers comprise the world's largest body of ice outside the polar ice caps. According to an inventory by the International Centre for Integrated Mountain Development (ICIMOD), the Indus is by far the most heavily glaciated of the region's main basins. Its 18,495 glaciers cover 21,193 km^2 and contain an estimated 2,696 km^3 of ice, representing 44 % of the total ice reserves in the entire HKH region (Bajracharya and Shrestha 2011). Glaciers act as massive regional freshwater repositories, seasonally accumulating snow and ice at high altitudes and releasing meltwater that contributes an estimated 35–50 % or more of the Indus' total flow (ICIMOD 2012; Savoskul and Smakhtin 2013). Snow and glacier melt constitute an especially critical water source during the shoulder seasons before and after the summer monsoon rains. In years of feeble or failed monsoons, meltwater can help avert or alleviate otherwise calamitous drought.

As climate change pushes up temperatures and shuffles precipitation patterns around the globe, glaciers worldwide are shrinking at unprecedented rates (Vaughan et al. 2013; Zemp et al. 2015).

Evaluations of the Indus find glacial area declining and glacier mass dwindling across most of the basin. Certain areas of the western Karakoram Range remain the noteworthy exception. Glaciers in this corner of northern Pakistan evince stability or slight growth (Bolch et al. 2012; Kääb et al. 2012; Kulkarni and Karyakarte 2014). All told, however, recent analyses estimate the Indus Basin's glaciers shed 7 gigatons (7 billion metric tonnes) of ice per year in 2003–2008 (Kääb et al. 2015).

Initially, increased glacier melting could boost river flows. As deglaciation continues, though, meltwater flows will subsequently wane, diminishing the downstream supplies available for drinking, sanitation, agriculture, hydropower, industry, and ecosystems (Singh et al. 2011; National Research Council 2012). Some research suggests that glacier meltwater patterns are already changing, shifting earlier in the year and contributing 3–13 % less water to stream flows across different catchments of the Indus (Mukhopadhyay 2012; Savoskul and Smakhtin 2013; Mukhopadhyay and Khan 2015).

Far from the mountain headwaters, climate change also jeopardizes freshwater resources in the Indus' coastal delta. As the planet warms, the oceans have absorbed 93 % of this additional heat over the past 50 years. The warmer ocean water in turn expands, like mercury in a thermometer, lifting sea levels worldwide. So too, the increased melting of mountain glaciers and land-based polar ice caps adds to river discharge flowing to the oceans, further boosting sea levels. Ice melt from continental glaciers outside Antarctica and Greenland accounts for about one third of the sea level rise observed since 1971 (Church et al. 2013). By one estimate, the shrinking glaciers of the Indus region in western South Asia could contribute 4.5–7.1 mm to global sea level rise by 2100 (Radic et al. 2014). All told, global mean sea levels have risen 19 cm since 1901 (Rhein et al. 2013). In the Indus Delta, climbing ocean levels cause seawater to seep into coastal aquifers, tainting local groundwater supplies. Similarly, mounting sea levels combined with declining river flows have allowed saltwater currents to creep up the river channel, reaching up to 225 km inland (Rasul et al. 2012; Syvitski et al. 2013). Saltwater intrusion compromises freshwater resources and ecosystems throughout the delta, contaminating agricultural land, debilitating fisheries, and damaging the mangrove forests that protect the coast from storm surges and erosion. Pakistan's Revenue Board figures that brackish seawater threatens more than 1.2 million acres of the delta (Anwar et al. 2014).

6.2.2 *Projected Climate Impacts*

The Indus Basin spreads over 1.12 million km^2 – an area equal to all of France, Germany, and Great Britain combined – encompassing multiple climatic zones ranging from subtropical arid plains in the south to snow-capped mountain highlands in the north (AQUASTAT 2011). Covering such a large territory, projected climate change impacts vary significantly across the basin. Deriving a full picture of global warming challenges in the Indus requires examining both broad regional and smaller subregional pressures.

To develop regional projections, the IPCC compared 42 different global climate models using a set of common scenarios for possible future concentrations of greenhouse gases (GHGs). Adopting a mid-range scenario that supposes world GHG emissions will peak around the year 2040 and decline thereafter, the IPCC projects mean annual temperatures in South Asia will warm 0.8–2.5 °C above present levels by mid-century (2046–2065). Projected precipitation changes range from −2 % to +26 %, with a middle estimate of +7 % (Meinshausen et al. 2011; Christensen et al. 2013).

Policymakers at all levels will need more fine-grained information to sustainably manage the Indus' waters. Experts from ICIMOD, the Indian Institute of Tropical Meteorology, and Nepal's Tribhuvan University have developed climate projections specifically for the Indus Basin (Rajbhandari et al. 2015); please see Appendix. Their modeling anticipates substantial warming throughout the region and throughout the year. The greatest warming occurs in the north, in the glaciated upper basin. Both monthly mean minimum and monthly mean maximum temperatures climb steadily over the twenty-first century. By the 2080s, winter minimum temperatures are 4 °C higher and summer maximum temperatures are 3.4–4.6 °C hotter than the historical averages (1961–1990). The models also find important changes in temperature extremes. By the 2080s, both the projected highest maximum temperatures and the lowest minimum temperatures spike by a whopping 4–8 °C over much of the basin.

Precipitation patterns change, too, though the spatial and seasonal tendencies are less distinct. Three different versions of the model simulations all show monsoon precipitation surging over virtually the entire basin by 2040, from 10 % higher in the central basin up to 50 % higher in parts of the north and east. In the second half of the century, though, two of the models find the increases persist in the east, but project monsoon precipitation sliding 10–20 % or more below 1961–1990 levels in the far north and much of the lower basin. All three model versions foresee winter precipitation will tumble in the lower basin by the 2040s, but grow in the upper Indus. Likewise, all three models expect increasing rainfall intensity (measured as millimeters of rain per day) for the northern and eastern basin through the end of the century, but declining intensity in the west.

Studies focused specifically on the Upper Indus Basin, stretching across northern Pakistan, India, and western China, largely parallel these results. A multi-model analysis of the upper basin, employing the same mid-range GHG concentrations scenario as the IPCC, projects mean maximum summer temperatures will warm 1.2 °C by mid-century and 1.9 °C by 2100, while mean minimum winter temperatures rise 1.9 °C in the period 2041–2070 and jump 2.4 °C by century's close. Re-running the analysis using higher GHG concentration assumptions, where emissions continue to swell through 2100, doubled these warming rates. Under both GHG concentration scenarios, annual average precipitation in the upper Indus increases 13–14 % by the 2040s and 13–23 % by the 2070s (Ali et al. 2015).

6.3 Climate Risks to Water Security

Water is a renewable resource, but freshwater supplies are not unlimited. Rain, snow and ice melt, seepage between surface waters and groundwater, and return flows from irrigation and other uses replenish rivers and lakes and recharge aquifers. For any given source, however, renewals vary over time and place. River flows and lake levels wax and wane through wet and dry seasons. Every watershed is only replenished by a finite amount of renewable water every year. The Indus Basin is already straining the limits of its allowance. Long-term available renewable water resources in the basin average 287 km^3 annually, including both surface and groundwater.

Against this supply, total annual water demand from the basin in the two main riparian states, India and Pakistan, ranges from 257 to 299 km^3 (Indus Basin Working Group 2013).

As water withdrawals increasingly outpace natural rates of renewal, the Indus has become what hydrologists call a "closed" basin (Smakhtin 2008; Molle et al. 2010). All of its available renewable resources are already allocated to various human and environmental needs, such that supply falls short of demand during part or all of the year. With human claims monopolizing the Indus' resources, scant water remains to sustain natural ecosystems. Since the 1990s, annual outflows to the delta have fallen under 12 km^3, far below the 25 % of mean annual runoff analysts think necessary to maintain vital habitats and ecological functions (Smakhtin et al. 2004; Kravtsova et al. 2009). At times the Indus no longer reaches the sea year round. With little or no spare capacity, closed basins have meager ability to accommodate new users or meet rising water demands. Yet growing populations and growing economies will drive up water requirements dramatically across the region in the coming decades.

Analyses by the International Water Management Institute project that India's domestic and industrial water withdrawals from the Indus will double over year 2000 levels by 2025 (Sharma et al. 2008). Municipal and industrial water use in Pakistan will nearly triple between 2010 and 2025 (Qureshi 2011). Agriculture, though, claims the lion's share of the basin's water, swallowing 278 km^3 of demand, or 93 % of total withdrawals (AQUASTAT 2011). The International Food Policy Research Institute anticipates India's irrigation water use on the Indus will climb 12 % from 1995 to 2025 (Ringler et al. 2009). By that same year, experts calculate Pakistan alone will need 250 km^3 of water to irrigate its fields (Qureshi 2011). On current trends, the Indus will not meet such demands. Assuming present policy regimes continue and existing levels of efficiency and productivity persist, models developed by the 2030 Water Resources Group comparing expected future water requirements against actually accessible, sustainable water sources project a 52 % gap between renewable supplies and annual demands on the Indian banks of the Indus in 2030. Pakistan will suffer nearly a 50 % shortfall (2030 Water Resources Group 2009).

6.3.1 Climate Change and the Water-Food-Energy Nexus

Climate change threatens to multiply the water stresses confronting the region, upsetting the amount, location, and availability of supplies even as policymakers strive to fulfill increasing needs and reconcile competing demands. Further complicating the resource governance challenges, vital water uses are themselves interconnected. Water managers characterize these interdependencies as constituting the water-food-energy nexus (Asian Development Bank 2013; FAO 2014). It takes water to produce food and energy. It takes energy to supply food and water. Water represents an essential input for agriculture, fisheries, and food supply chains. It is used extensively in energy generation, for hydropower, cooling thermal power plants, and mining. Likewise, growing, preparing, preserving, and distributing food require energy, as do extracting, pumping, transporting, and treating water. In Pakistan's Punjab province, for example, agriculture consumes 20 % of total energy demand, while pumping irrigation water absorbs 61 % of all on-farm energy use (Siddiqui and Wescoat 2013). Agricultural practices – what crops to grow, how, and where – substantially affect local water cycles. Many crops can be turned to energy as biofuels. Policy choices at various points in the water-food-energy nexus can have synergistic or conflicting impacts at other points. Large-scale infrastructure like a dam may provide hydroelectricity and water storage for cities upstream but potentially disrupt downstream agriculture and ecosystems.

Global warming risks inflicting substantial strains on the water-food-energy nexus across the Indus Basin. Consider food. Climbing temperatures and more variable precipitation would generally increase water demand for crop irrigation (Bates et al. 2008). Initially, higher stream flows from greater monsoon rainfall and glacial melt might help fill this larger demand. Over time, however, as meltwater contributions to the Indus decline, one study found mean water supply decreasing by 8.4 % by mid-century. When integrated with assessments of irrigation requirements and crop yields, such shifts suggest the basin's water resources would be able to feed 26 million fewer people in 2050 than today even as the region's population will then be considerably higher (Immerzeel et al. 2010). By the end of the twenty-first century, multiple models conclude future water demands on the Indus will exceed climate constrained supplies, with consumers depleting groundwater reserves at unsustainable rates to meet their needs (Hejazi et al. 2014; Wada and Bierkens 2014).

Looking more closely at individual countries, the World Bank has modeled agricultural impacts in Pakistan over an 80-year horizon (Yu et al. 2013). Analyses of 70 plausible "climate futures" project overall crop production would decrease some 3 % on average, cutting GDP by 1.1 % and lessening household incomes by 2 % on an annual basis. In the worst-case climate scenario considered, crop production tumbles 13 %, GDP slides 2.7 %, and household incomes plunge 5.3 %. In India, the Indus Basin is the nation's breadbasket. Punjab alone produces 20–22 % of the country's wheat and 10–12 % of its rice and provides 50–75 % of the wheat and 30–48 % of the rice for the national buffer stock of food grains (Dhillon et al. 2010). Climate change menaces both staples. Higher temperatures impact crop yields. Studies in Punjab figured grain yields slump 3–10 % for rice and 10–18 % for wheat for temperature increases of 1–2 °C compared to normal conditions (Chauhan et al. 2014). Rice, a *kharif*, or monsoon season, crop could be especially vulnerable to swings between extreme surplus and deficit monsoons and to potential drops in monsoon rainfall later this century. Modeling analyses suggest a 10 % deficit below average rainfall diminishes rice production 11 %. A 30 % rainfall deficit slashes production 32 % (Praduman et al. 2014).

To enhance the resilience of their water-food-energy systems, all four Indus riparian states intend to significantly augment both their water storage and hydropower capacities (Department of Climate Change 2015; Government of India 2015; Islamic Republic of Afghanistan 2015; Ministry of Climate Change 2015a). Afghanistan currently possesses 4 hydroelectric dams on the Kabul and plans 11 more (AQUASTAT 2013). China has constructed a dam on the Indus and one on the

Sutlej (Chaturvedi 2013, p. 184). India maintains 6 large hydropower dams on the Beas, Chenab, and Sutlej, as well as 4 major irrigation barrages, while Pakistan has 3 major hydro dams on the Indus and Jhelum, plus 15 irrigation barrages (Indus Basin Working Group 2013).

Yet climate change could considerably affect hydropower and other infrastructure schemes across the Indus. If peak snow and ice meltwater contributions to the river decline or move earlier in the year, or if monsoon rains fail more frequently or dramatically, water storage for running turbines or feeding irrigation will have to last longer through dry seasons and droughts. Shifts in the timing, location, and quantity of river flows may mean there will not be enough – or will be too much – water where and when it is needed. A 1 % drop in stream flow, for example, can trim hydropower production by 3 % (Laghari 2013). Fluctuating stream flows may also exacerbate erosion and sedimentation. Sedimentation can raise the riverbed, rendering it more susceptible to flooding, and accumulate in reservoirs, reducing their flood control and storage capacity. The many uncertainties surrounding climate projections, however, confound ready policy planning. Examining global warming's potential impacts on Pakistan, the World Bank calculated hydropower generation could soar up to 22 % or plummet as much as 34 %, depending upon the climate scenario (Yu et al. 2013).

Extreme climate events pose further dangers to the water-food-energy nexus. The record Pakistan floods of 2010 killed 1,600 people, displaced 20 million, and inundated 38,000 km^2 of the country. Of the $10 billion in total damages, 50 % struck the agricultural sector. Energy, water supply and sanitation, and flood control and irrigation accounted for another 7 % (Ali 2013). Most of the destruction was caused at the intersections of the water-food-energy nexus as floodwaters backed up and overflowed behind dams and barrages or surged through failed irrigation levees and breached barrages (Syvitski and Brakenridge 2013). To be sure, no one natural disaster can be attributed to global warming. Even so, sophisticated statistical analyses indicate climate change is already increasing the occurrence of extreme temperatures and precipitation (Hansen et al. 2012; Hao et al. 2013; Fischer and Knutti 2015). Similar studies suggest global warming will drive more frequent and severe flooding and droughts in the Indus Basin's future (Fischer et al. 2013; Dankers et al. 2014; Kundzewicz et al. 2014).

The upper Indus is also vulnerable to a particular disaster risk specific to glaciated watersheds. As climate change shrinks mountain glaciers, meltwater often collects in pools behind natural barriers of ice or debris. Increasing water pressure as the lake fills, seismic activity, avalanches, landslides, or other triggers can weaken or collapse these retaining walls, sending sudden waves of water rushing downriver as "glacial lake outburst floods" (GLOFs). Such floods have reached more than 1,200 km downstream in the upper Indus and can cause significant loss of life and damage to roads, bridges, and other infrastructure. In 1985 an upstream GLOF demolished the Dig Tsho hydropower installation in Nepal 2 weeks before its inauguration (Reynolds 2014). In the Indus Basin, ICIMOD has catalogued 16 potentially dangerous glacial lakes on the Indian tributaries and a further 52 in Pakistan (Ives et al. 2010).

6.3.2 Climate Resilience and Water Security in Transboundary Basins

How can decision-makers negotiate the water-food-energy nexus in a warming world? Water policy is inherently complex, linking multiple sectors and stakeholders over different geographic scales and timeframes. Rarely do administrative boundaries and hydrological borders coincide. Environmental pressures increasingly impose novel water security challenges, undermining standard assumptions about the variability of natural systems and altering parameters such as precipitation, river discharge, and probabilities of extreme events that have long governed water planning and risk assessment (Milly et al. 2008). Growing recognition of the interdependent and multidimensional character of water governance has forged new management paradigms to promote more coordinated management of water and related resources. These policies take various names – integrated water resources management (IWRM), adaptive water management, nexus approaches, ecosystem-based strategies – and differ in selected particulars and emphases, but all share the basic objectives of holistic and sustainable multilevel water governance (Schoeman et al. 2014; Benson et al. 2015).

No one management model can apply uniformly to all basins worldwide. Apt solutions must be tailored to specific places, polities, and policy contexts. Nevertheless, certain central principles emerge from the evolving integrative governance paradigms (INBO 2013; OECD 2015). First, policymaking should be participatory and transparent to ensure public legitimacy. To secure the commitment and contribution of relevant actors, and to navigate trade-offs among competing users and contending demands, authorities should promote stakeholder dialogue and input to policy design and implementation. Second, policy must be scientifically informed and evidence-based to be effective. Authorities should produce and share timely, consistent, and comparable water data and information to guide, evaluate, and improve resource management. Third, policy must be adaptable and adjustable to meet the complexities of the water-food-energy nexus and manage climate threats that will evolve in uncertain ways over lengthy timeframes. Authorities should embrace learning by doing, iteratively assessing policy impacts, incorporating new information and experience, and revising implementation accordingly. Most importantly, policymaking must recognize the basin as a hydrological unit and manage water at functionally appropriate scales within integrated basin governance systems. Where political or sectoral boundaries nest within or intersect at various scales across the basin, management practices should foster multilevel cooperation and cross-sectoral coordination among users and levels of government. Cooperative knowledge building, stakeholder dialogue, and integrative multilevel cooperation provide the key to managing shared climate challenges in transboundary basins where adaptation options pursued by one nation can impact the water resource risks perceived by others and separate resilience strategies may prove insufficient to assure the sustainability of interdependent water-food-energy systems (Sadoff and Muller 2009; WWAP 2012; Rieu-Clarke et al. 2015; UNECE 2015).

6.4 Climate Change and Water Policies in the Indus Basin States

Policymakers in the Indus Basin states recognize the mounting pressures on water resources posed by global climate change. The national climate strategies – termed Intended Nationally Determined Contributions (INDCs) – prepared by Afghanistan, China, India, and Pakistan in advance of the landmark 21st Conference of the Parties to the United Nations Framework Convention on Climate Change (UNFCCC), gathered in Paris in 2015 to fashion an international agreement to combat global warming, all emphasize the climate threat to freshwater supplies, services, and infrastructure (Department of Climate Change 2015; Government of India 2015; Islamic Republic of Afghanistan 2015; Ministry of Climate Change 2015a). Publics in the riparian nations are also increasingly cognizant of the emerging climate dangers. Recent polling revealed sizable majorities of Chinese, Indians, and Pakistanis alike concerned that global warming will impact them personally within their lifetimes [Afghanistan was not surveyed]. Tellingly, drought and water shortages topped the list of Chinese and Indian climate worries, while Pakistanis most fear flooding and extreme weather (Pew Research Center 2015).

Beyond the shared political and public awareness of gathering climate risks, the Indus nations bring disparate capacities and priorities to the tasks of strengthening sustainable water management and building climate resilience. In India and Pakistan, the basin's water stores are increasingly stretched by the contending needs of hydropower, soaring urban populations, and supplying the most intensively irrigated agricultural areas on earth. In Afghanistan and China, continuing political turmoil in the first case and forbiddingly remote geography in the second have inhibited developing river resources, though demands are rising. Yet the many common climate threats and policy challenges identified by the basin states make clear the multiple opportunities and manifest mutual benefits of greater cooperation and exchange to manage their collective water resources.

6.4.1 Afghanistan

Ensuring the nation's water security constitutes a cornerstone of Afghanistan's climate resilience strategies. Both Afghanistan's Initial National Communication (INC) to the UNFCCC, detailing its climate vulnerabilities and policies, and its National Adaptation Programme of Action for Climate Change (NAPA) single out water resource risks – droughts, floods, and consequent threats to crops, livestock, and agricultural livelihoods, etc. – as the most critical hazards facing the country. Across 11 main adaptation programs evaluated for preserving human life and health, protecting vulnerable populations, and enhancing poverty reduction and food security, the NAPA and INC place greatest priority on two areas: increasing agricultural water use efficiency and improved land and water management at the watershed level. Afghanistan's National Water Law and Water Sector Policy share this outlook, embracing integrated water resources management at the basin and sub-basin scale. Similarly, Afghanistan's National Disaster Management Act and Strategic National Plan for Disaster Risk Reduction seek to curb future natural disaster risks by proactively incorporating climate adaptation approaches into disaster management policy (National Environmental Protection Agency 2009, 2012).

Nevertheless, Afghan authorities acknowledge confronting severe capacity constraints. Afghanistan's INC underscores the paucity of systematic and reliable data and research on climatological and hydrological phenomena at policy-relevant scales to inform decision-makers. Inadequate financial and technical resources further hobble effective policy. Kabul calculates the country faces funding gaps of $7.85 billion to strengthen hydrometeorological monitoring, rehabilitate infrastructure, plan watershed management, and increase irrigated agriculture. Recognizing the scale of these challenges, Afghan climate policy explicitly calls for enhanced international cooperation, including sharing best practices for climate assessments and adaptation, forming university partnerships for joint research, and coordinating with regional organizations on the transfer of environmentally sound technologies (National Environmental Protection Agency 2012; Islamic Republic of Afghanistan 2015).

6.4.2 China

As China's population has grown and its economy industrialized in recent decades, the twin challenges of water scarcity and water pollution have spread. In response, the government promulgated policies of strictest water resources management (SWRM), embodied in the so-called three red lines: control of water development and utilization, control of water use efficiency, and control of pollutants. SWRM further aims to fix annually binding targets on agricultural, industrial, and drinking water withdrawals, limiting total national water use to 700 km^3 by 2030 (Zuo et al. 2014).

Global warming imperils China's sustainable water management goals. Beijing's Second National Communication to the UNFCCC affirms that "changing climate has already changed distribution of water resources in China, and its impacts will be even more significant in the future" (National Development and Reform Commission 2012, p. 12). In response, the government's National Strategy for Climate Change Adaptation specifies multiple measures to foster water sector resilience. China has established new meteorological disaster screening and early warning systems, completed a national water census and technical reviews of water allocation for 25 major basins, operationalized a national drought control plan, and launched large-scale projects to bolster water efficiency in irrigation and power generation (National Development and Reform Commission 2014).

Reflecting its position as the world's most populous country, largest GHG emitter, and, measured by gross national income at purchasing power parity, now its biggest economy (World Bank 2015), China has increasingly assumed a leadership role in global climate policy. Though careful to couch its international commitments within the context of its national capabilities, China has pledged to create a fund for South-South Cooperation on Climate Change and to promote mutual learning, policy coordination, and sharing of good practices between developing countries (Department of Climate Change 2015).

6.4.3 India

India's INDC "identifies water as the most critical component of [the] life support system" (Government of India 2015, p. 21). Water accordingly figures as one of the eight national missions comprising India's National Action Plan on Climate Change. The National Water Mission aims to enhance integrated water resources management, ensure equitable resource allocation between and within the states, promote water use efficiency and conservation, augment surface and groundwater storage capacities to address variability in precipitation and river flows, and expand irrigation and water reuse and recycling. Separate national missions devoted to the Himalayan ecosystem and to sustainable agriculture contain components to protect the Himalayan sources and adapt the agricultural uses of India's water supplies (Prime Minister's Council on Climate Change 2008).

In parallel, India's 2012 National Water Policy establishes basin-scale IWRM as the guiding principle for planning, developing, and governing water resources (Ministry of Water Resources 2012). Condemning the human and ecological costs of inefficient policies and wasteful practices, the policy warns that absent effective management, rising demands, deteriorating pollution, and growing climate pressures will increasingly strain scarce water supplies, potentially engendering conflicts between different users. Climate impacts should thus be factored into all water decision-making, including disaster preparedness and response, and the planning and operation of infrastructure such as dams, reservoirs, irrigation, and industrial projects should include climate coping strategies.

The National Water Policy accompanies its argument for adopting interdisciplinary and integrated management strategies with an emphasis on erecting transparent, participatory, and cooperative institutional frameworks engaging all the relevant states, sectors, and stakeholders. On transboundary rivers, it calls for forging international agreements on real-time data exchange and negotiations on water sharing and management with other riparian nations, though it carefully stipulates such arrangements should be reached on a bilateral basis.

6.4.4 Pakistan

Pakistan's 2012 National Climate Change Policy (NCCP) announces several key objectives including integrating climate change into related national policies; ensuring water, food, and energy security; and minimizing natural disaster risks (Ministry of Climate Change 2012). Detailing the dangers of global warming, it puts threats to water resources at the top of Pakistan's climate vulnerabilities. Increased monsoon variability and retreating Himalayan glaciers jeopardize Indus flows. Worsening flood and drought hazards and saline intrusion into the Indus Delta menace water-dependent communities and ecosystems. And rising water stresses could heighten tensions between upper and lower riparian regions over water sharing.

Presenting Pakistan's climate adaptation measures, the NCCP highlights strategies to sustain the water-food-energy nexus, citing plans to expand hydropower production, augment water storage capacities, improve irrigation efficiency, restore environmental flows to the delta, and apply IWRM to water allocations and groundwater management. The NCCP also advances specific possibilities for international cooperation. It suggests exploring joint management of transboundary watersheds with neighboring countries, proposes entering a water treaty with Afghanistan, considers forming agreements among regional states to declare the Himalayan glaciers a "protected area," and advocates regional sharing of hydrological and meteorological data and strengthening coordination mechanisms among international water sector institutions.

Even so, Pakistan's climate and water policies appear in transition. In September 2015, the Ministry of Climate Change released a "zero draft" of Pakistan's INDC, confirming and reiterating the NCCP in preparation for the Paris conference (Ministry of Climate Change 2015a). To the dismay of many observers, however, the paper was never delivered to the UNFCCC, replaced instead by a single-page submission mentioning neither particular vulnerabilities nor specific adaptations (Ministry of Climate Change 2015b; Ebrahim 2015). Pakistan's water governance is also in flux, with Islamabad expected to finalize a new national policy by the end of 2015 (Haider 2015).

6.5 Opportunities for Climate Cooperation

The Indus watershed spreads across four sovereign states. Policymaking in the basin is thus divided among multiple different national and subnational actors, institutions, and stakeholders, often pursuing distinct interests and serving disparate constituencies. Yet all the Indus nations now face complex and intertwined climate pressures and water security risks. Greater dialogue and coordination among the basin's diverse communities offers considerable scope for mitigating mutual threats and generating collective benefits. Collaborative approaches promoting joint monitoring and technical exchange, data sharing, and knowledge generation can help policymakers and the broader public better apprehend and assess the basin's complex climate and water challenges. Cooperative disaster preparedness and response can build regional resilience against future risks of worsening climate impacts. Common frameworks for collecting and disseminating policy lessons can enlarge the range of policy choices, scale up best practices, and chart potential pathways forward. The national policies of the riparian countries espouse such cooperative aims. Institutional structures exist to help enable collaborative strategies. Policymakers must enact them.

6.5.1 Information

Policymakers in the Indus Basin lack consistent, systematic hydrological and climatological data necessary to effectively pursue integrated water resources management. Data collection and observation networks are inadequate and their coverage fragmented, typically focusing on single countries. Research efforts are largely uncoordinated between riparian nations or even among institutions within the Indus states, and their results insufficiently disseminated. Scientific analyses are too often unconnected to policymaking needs and policymaking divorced from scientific inputs (Karki et al. 2011; Cheema and Pawar 2015). Without sound information and understandings of environmental and socioeconomic conditions, drivers, and trends affecting water resources, decision-makers cannot accurately appraise potential climate change impacts or judiciously design and implement adaptation measures and resilience policies.

Increased data sharing and cooperative research would allow the Indus nations to more efficiently pool their respective information resources to develop a more robust and comprehensive knowledge base for policy formation. Regular data exchanges, joint monitoring, and coordinated research programs would help strengthen national and regional scientific capacities, contribute to establishing common understandings of shared climate and water resource risks, and build trust among policy and stakeholder communities around the region. Similarly, given that the basin states confront many of the same policy challenges, information exchanges on policy initiatives, sharing policy lessons, and cooperative policy reviews and assessments could allow the riparian states to identify and spread successful approaches and scale up best practices.

Key fields for cooperative knowledge building include expanding and upgrading the network of hydrometeorological monitoring stations; sharing precipitation, stream flow, and water quality data; tracking monsoon trends and variability; utilizing remote sensing technologies and in situ observation to improve understanding of glacier behavior and dynamics; developing down-scaled regional climate impact models; and better mapping and characterizing the basin's groundwater aquifers (Karki et al. 2011; Cheema and Pawar 2015).

Three particular areas merit special attention. First, agriculture commands the bulk of water use in the Indus. But much of it is wasteful. Irrigation efficiency in the Pakistani portion of the Indus hovers around 40 %, while across India canal irrigation efficiency stands at about 38–40 % (Frenken 2012). Considerable opportunities therefore exist for sharing policy lessons and collaborative evaluation and adaptation of best practices in agricultural water management. Cooperative research and agricultural extension could promote the deployment of such water-saving techniques as laser land leveling, drip irrigation, wastewater recycling, improved crop planning and diversification, and water pricing (Erenstein 2009; Laghari et al. 2012).

Second, the Indus nations are acutely vulnerable to natural disasters. Such catastrophes often transcend national borders. Advances in monitoring and communications technologies now enable continuous collection and real-time transmission of hydrometeorological data. Observation stations can record weather variables on SIM cards, for example, capturing spiking temperatures, rapidly rising rainfall, surging stream flows, etc., and relay the data over the Internet. Capitalizing on these capabilities, increased information sharing and technical and professional cooperation between the riparian states could significantly advance national and regional flood forecasting and disaster early warning systems (Shrestha et al. 2015; Zia and Wagner 2015).

Finally, the lack of transparent information feeds popular apprehensions and political acrimony. Indus neighbors often suspect one another's actions and mistrust one another's intentions (Price et al. 2014). Defusing these tensions calls scientists and researchers not only to exchange information but to raise public awareness and inform public debate. The region's expert communities must develop cooperative communications strategies such as coordinated media campaigns, joint parliamentary briefings, and collaborative analyses by institutions representing the different Indus nations working together to cultivate collective knowledge and shared understandings of the basin's common challenges and opportunities (Indus Basin Working Group 2013).

6.5.2 Infrastructure

Water infrastructure – reservoirs, dams, irrigation works, sanitation systems, etc. – allows humans to mitigate water risks and generate water benefits (Grey and Sadoff 2007; Briscoe 2009). Yet the Indus states have not developed the infrastructure to manage their water resources as fully as they might. India, for instance, has built or begun construction on 48 % of an estimated 33,832 MW of hydropower capacity on its side of the basin. Pakistan has developed just 11 % of a potential 59,208 MW (Indus Basin Working Group 2013). The basin states also possess scant water storage capacity by international standards. Compared with an OECD average of 2,291.3 m^3 of water storage per capita, Afghanistan has 67.4 m^3 of water storage per person, China's reservoirs stock 412.1 m^3 per capita, and India enjoys 193.3 m^3 of water storage per head. Holding 155.2 m^3 of water per inhabitant, all Pakistan's reservoirs combined store just 30 days of supply for the entire country (AQUASTAT 2015).

Infrastructure development for increased hydropower and water storage figures prominently in all the Indus countries' climate adaptation strategies. It also ties the riparian states together commercially and politically. China's "One Belt, One Road" initiative will fund hydropower projects in Pakistan (Xinhua 2015). India is assisting Afghanistan to erect 12 dams on the Kabul River (Shroder 2014, p. 507). But infrastructure developments have proven highly controversial both between and within the basin states (Kugelman and Hathaway 2009; Qiu 2012; Akhter 2013; Price et al. 2014). Downstream communities fear that dams, reservoirs, and irrigation schemes constructed upstream will decrease or disrupt vital water flows downriver. Upstream communities worry that caveats and conditions posed by downstream neighbors would curtail possibilities to exploit the river's resources. Many observers in the Indus and beyond further fault large infrastructure projects on several grounds. Decision processes for funding, siting, building, and operating projects can be opaque, subject to biases and corruption. Costs and benefits are often unequally distributed, disfavoring marginal populations. Negative social and environmental consequences – which may outweigh the anticipated returns – are often downplayed or disregarded, if they are appropriately assessed and incorporated into decision-making at all (Duflo and Pande 2007; Richter et al. 2010; Beck et al. 2012; Ansar et al. 2014).

For these very reasons, infrastructure programs in the Indus must be carefully evaluated from an integrated management perspective to realize their objectives and reconcile trade-offs. So-called "run-of-the-river" hydropower dams, for example, may assuage downstream concerns over potentially diminished stream flows because they don't impound large reservoirs. Lacking significant water storage capacity, however, these same facilities are consequently vulnerable to droughts, floods, and fluctuating river flows following from climate change. Where negotiating such challenges raises transboundary implications, infrastructure projects must be developed transparently in consultation with the affected riparian states to be economically, environmentally, and politically sustainable.

To this end, the Indus nations can adopt several cooperative strategies (Indus Basin Working Group). The riparian countries should establish a professional exchange program for engineers and water managers from each country collaboratively to identify, expand, and implement agreed best practices for the sustainable management of hydroelectric facilities and other infrastructure, including sediment flushing, maintaining environmental flows, flood control protocols, and potential modalities for joint observation of operations. The riparian countries should further conduct and release joint studies of the cumulative environmental and socioeconomic impacts of multiple dams operating concurrently on single waterways, including long-term risk assessments in light of climate change. Both measures should support the elaboration and adoption of an internationally standardized environmental impact assessment for infrastructure on the Indus.

6.5.3 Institutions

The 1960 Indus Waters Treaty (IWT) between India and Pakistan represents the primary institutional framework for international water management on the Indus. In stark contrast to the subsequently formulated principles of IWRM, the Treaty physically divides the river. It allots unrestricted use of the waters of the three western tributaries – the Jhelum, Chenab, and Indus – to Pakistan. India receives the use of the three eastern tributaries – the Beas, Ravi, and Sutlej. The IWT contains no mechanisms for addressing the variations in river flow that climate change could engender, and no provisions regarding the groundwater resources that India and Pakistan also share. Neither does it include Afghanistan or China, which are not parties to the accord.

Even so, the IWT offers several foundations for building stronger transboundary cooperation. Critically, the Treaty mandates the regular exchange of information on river flows and water utilization or any other requested data and requires each country to inform the other if it undertakes engineering works that could impact the other party. Beyond information sharing, the IWT Article VII on "Future Cooperation" notably declares the parties' intention to collaborate to install hydrometeorological monitoring stations, carry out drainage works, and undertake joint engineering works. These terms offer a ready template for reciprocal data sharing and policy coordination that should be enhanced between India and Pakistan and could be extended to Afghanistan and China. The IWT itself makes no allowance for the adherence of additional parties, but such cooperation could be regularized through MoUs between the participants. China and India, for example, already maintain an MoU for flood season data sharing on the Sutlej (Ministry of Water Resources 2015). And indeed, Afghanistan and Pakistan have engaged discussions for a water-sharing agreement on the Kabul River and announced plans for a joint hydropower plant on the Kunar tributary (Vick 2014).

Numerous other institutional avenues exist to support enhanced transboundary cooperation. All four Indus states belong to ICIMOD, an intergovernmental research organization that furnishes a "neutral" platform for data collection, knowledge building, and joint assessment of policy lessons and best practices. Afghanistan, India, and Pakistan are likewise members of the South Asian Association for Regional Cooperation (SAARC), while China is an observer. Although questions do surround the organization's efficacy, SAARC members have adopted a convention on environmental cooperation calling for regional policy collaboration. SAARC has also established several research centers – including the SAARC Disaster Management Centre, a SAARC Meteorological Research Centre, the South Asia Forum, and the new South Asian University – that could carry out cooperative projects and embed evaluations of basin-wide water governance challenges in a multilateral setting.

Finally, the Indus nations should promote opportunities for capacity building and policy learning and collaboration at the subnational province/state, city-to-city, and civil society levels (Indus Basin Working Group 2013). Some neighboring jurisdictions in the basin have managed a modicum of local cooperation in other issue areas that might serve as a model for exploring subnational collaboration on the Indus. Pakistani and Indian Punjab, for instance, reached an MoU to boost cross-border trade, backed by local business communities (Maini 2012). Sharing many of the same water management challenges for providing municipal water and sanitation, city governments are especially well placed to share best practices, backed by growing support networks for city-to-city initiatives (Singh 2008).

6.6 Conclusion

Water security, "the availability of an acceptable quantity and quality of water for health, livelihoods, ecosystems, and production, coupled with an acceptable level of water-related risks to people, environments and economies," is essential to human welfare (Grey and Sadoff 2007, pp. 547–8). In the Indus Basin, water security is increasingly endangered. Surveys in Afghanistan show problems with water provision for drinking and irrigation sow popular dissatisfaction with public service and government performance (Warren and Hopkins 2015). In China, senior officials openly worry that worsening water stress has come to impede economic development and could fuel civil unrest (Economy 2011; Walker et al. 2014). In India and Pakistan, government and business leaders regularly warn that constraints on sustainable water supplies jeopardize economic growth and societal well-being (Planning Commission of Pakistan 2011; Hornby 2012; Perveen et al. 2012; Stancati 2012; Masood 2015). Global warming will aggravate water management challenges across the basin. A study of ten major Asian rivers concluded that the combination of mounting socioeconomic stresses, increasing environmental strains, and persistent governance shortcomings renders the Indus Basin most vulnerable of all to aggregate risk (Varis et al. 2012).

Seven thousand years ago, South Asia's earliest civilization rose on the banks of the Indus. Its towns and cities flourished from northern India and Pakistan to the Arabian Sea. Recent archeological investigations suggest that climatic shifts brought weakening monsoons and deepening drought that dried the rivers on which those urban centers relied, precipitating the ultimate collapse of Harappan society (Giosan et al. 2012; Dixit et al. 2014). Yet contemporary "Industan" is not condemned to suffer the fate of its Bronze Age predecessors. Collaborative knowledge building, international dialogue, and integrative multilevel policy cooperation provide the Indus nations practical strategies for effective and sustainable water governance. Whether the riparian communities can effectively achieve such a collective vision for managing the basin's shared resources remains an open question. Perhaps the only certainty is that, if they fail to cooperate together in the face of climate change, then together they will bear its consequences.

Appendix (Figs. 6.1, 6.2, 6.3, 6.4)

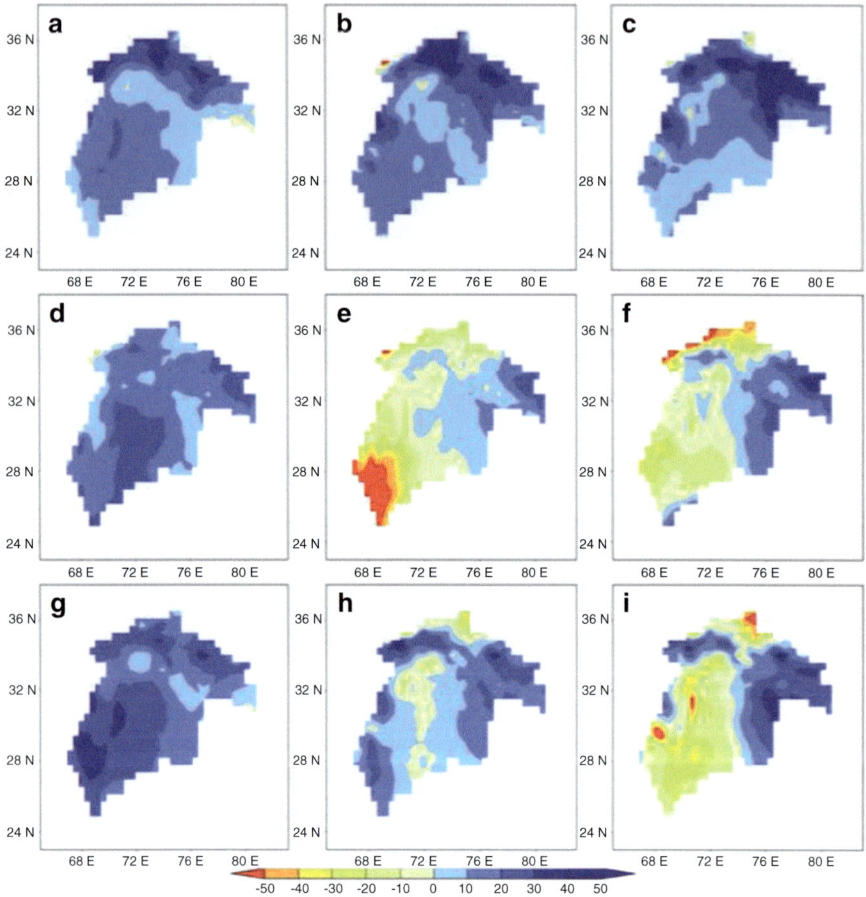

Fig. 6.1 Projected % change in mean summer monsoon rainfall with respect to baseline period 1961-1990 in 2011–2040 (*left column*), 2041–2070 (*middle column*), and 2071–2098 (*right column*), produced by the Providing REgional Climates for Impact Studies (PRECIS) regional climate modelling system using three different simulations from the Hadley Centre Met Office Quantifying Uncertainty in Modelling Projections (QUMP) project, Q0 (*top row*), Q1 (*middle row*), and Q14 (*bottom row*). The selected simulations are described in Rajbhandari et al. (2015). (Source: Rajbhandari, R. et al., "Projected Changes in Climate over the Indus River Basin Using a High Resolution Regional Climate Model (PRECIS)," Climate Dynamics 44, no.1 (2015), p.347)

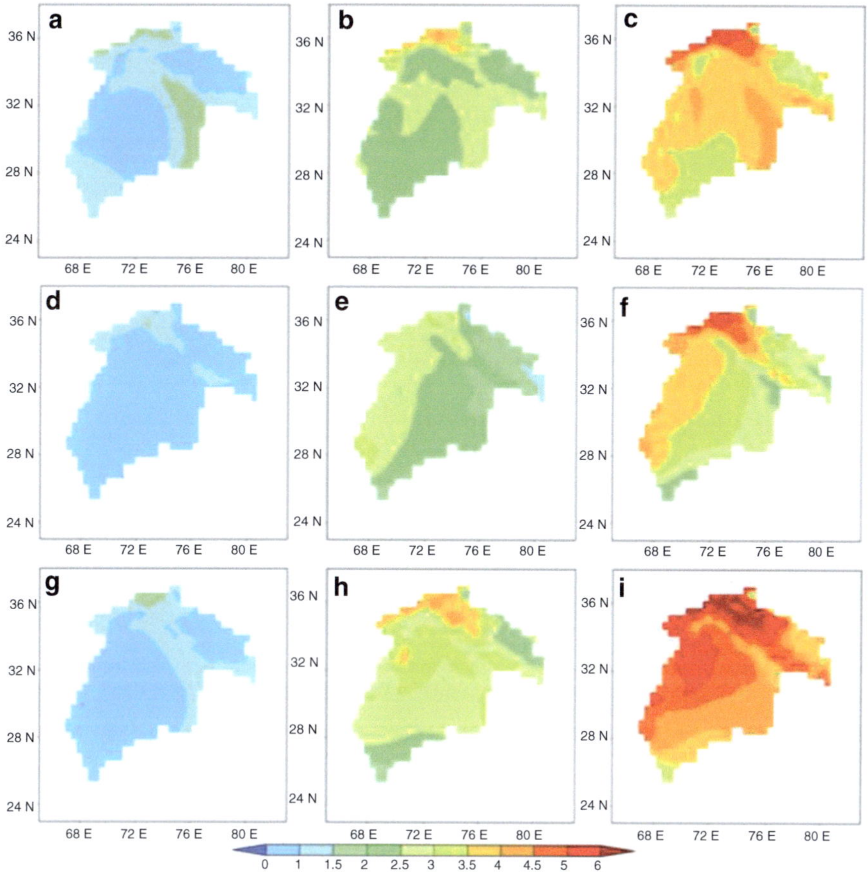

Fig. 6.2 Projected oC change in summer season maximum temperature with respect to baseline period 1961–1990 in 2011–2040 (*left column*), 2041–2070 (*middle column*), and 2071–2098 (*right column*), produced by the Providing REgional Climates for Impact Studies (PRECIS) regional climate modelling system using three different simulations from the Hadley Centre Met Office Quantifying Uncertainty in Modelling Projections (QUMP) project, Q0 (*top row*), Q1 (*middle row*), and Q14 (*bottom row*). The selected simulations are described in Rajbhandari et al. (2015). (Source: Rajbhandari, R. et al., "Projected Changes in Climate over the Indus River Basin Using a High Resolution Regional Climate Model (PRECIS)," Climate Dynamics 44, no.1 (2015), p.348)

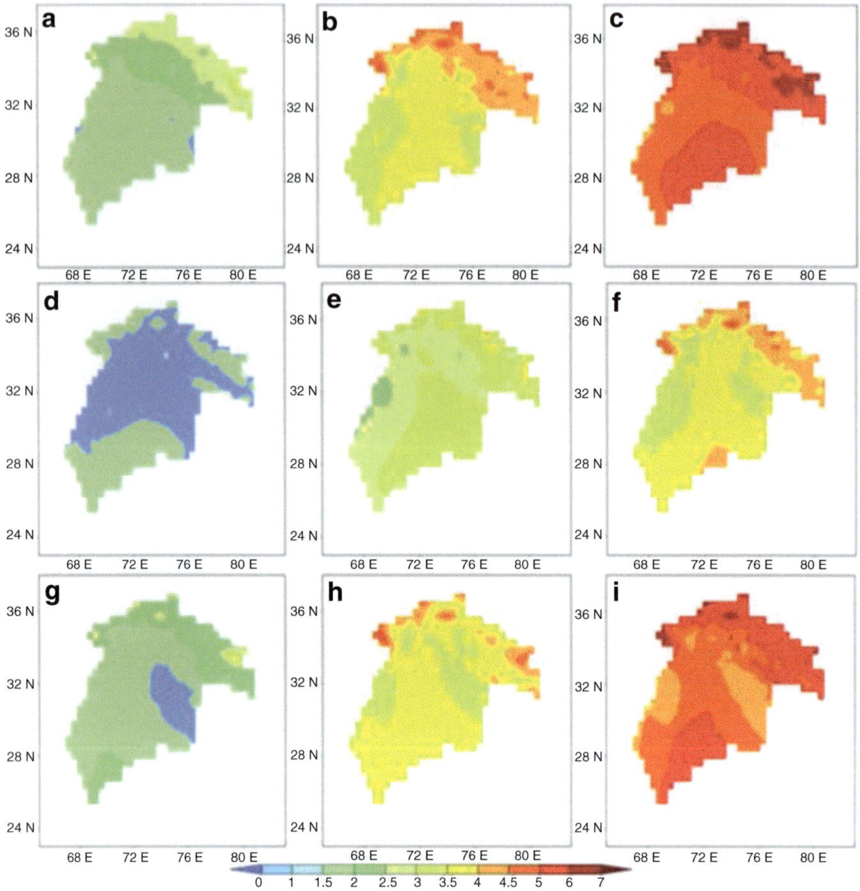

Fig. 6.3 Projected oC change in winter season minimum temperature with respect to baseline period 1961–1990 in 2011–2040 (*left column*), 2041–2070 (*middle column*), and 2071–2098 (*right column*), produced by the Providing REgional Climates for Impact Studies (PRECIS) regional climate modelling system using three different simulations from the Hadley Centre Met Office Quantifying Uncertainty in Modelling Projections (QUMP) project, Q0 (*top row*), Q1 (*middle row*), and Q14 (*bottom row*). The selected simulations are described in Rajbhandari R et al. (2015). (Source: Rajbhandari, R. et al., "Projected Changes in Climate over the Indus River Basin Using a High Resolution Regional Climate Model (PRECIS)," Climate Dynamics 44, no.1 (2015), p.349)

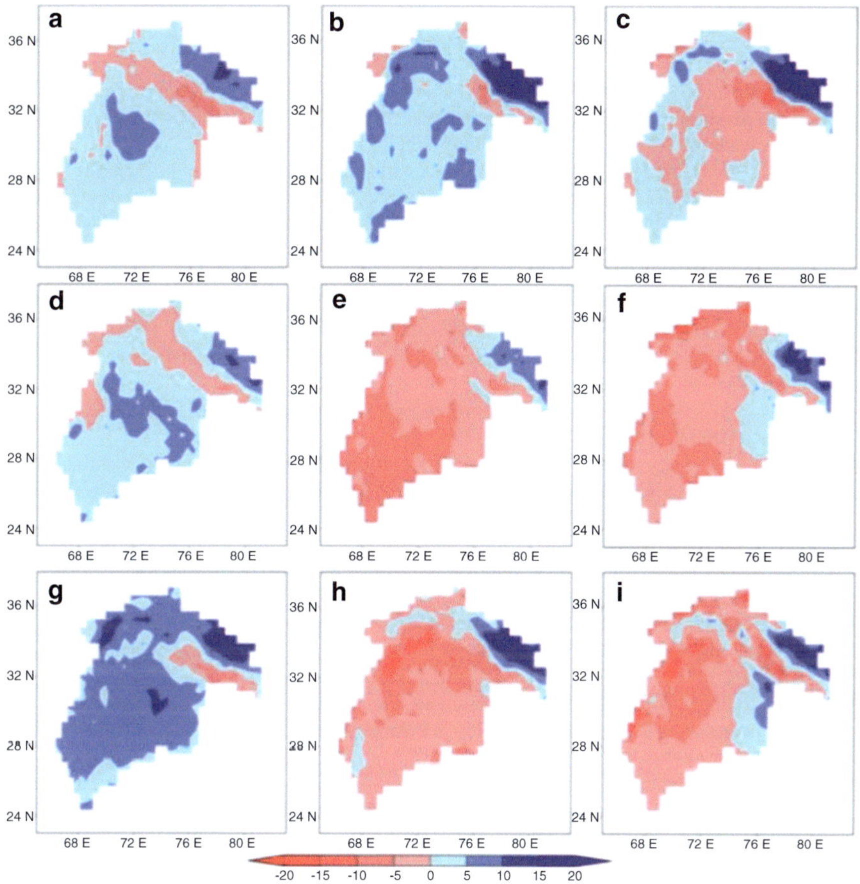

Fig. 6.4 Projected change in annual number of rainy days with respect to baseline period 1961-1990 in 2011–2040 (*left column*), 2041–2070 (*middle column*), and 2071–2098 (*right column*), produced by the Providing REgional Climates for Impact Studies (PRECIS) regional climate modelling system using three different simulations from the Hadley Centre Met Office Quantifying Uncertainty in Modelling Projections (QUMP) project, Q0 (*top row*), Q1 (*middle row*), and Q14 (*bottom row*). The selected simulations are described in Rajbhandari et al. (2015). (Source: Rajbhandari, R. et al., "Projected Changes in Climate over the Indus River Basin Using a High Resolution Regional Climate Model (PRECIS)," Climate Dynamics 44, no.1 (2015), p.351)

References

2030 Water Resources Group (2009) Charting our water future: economic frameworks to inform decision-making (McKinsey & Company). At: http://www.mckinsey.com/App_Media/Reports/Water/Charting_Our_Water_Future_Full_Report_001.pdf

Akhter M (2013) Geopolitics of dam design on the Indus. Econ Polit Wkly 48(19):24–26

Ali A (2013) Indus basin floods: mechanisms, impacts, and management. Asian Development Bank, Manila. At: http://www.adb.org/sites/default/files/publication/30431/indus-basin-floods.pdf

Ali S et al. (2015) Twenty first century climatic and hydrological changes over upper Indus Basin of Himalayan Region of Pakistan. Environ Res Lett 10 (014007). At: http://m.iopscience.iop.org/article/10.1088/1748-9326/10/1/014007

Ansar A et al (2014) Should we build more large dams? The actual costs of hydropower megaproject development. Energ Policy 69(43):43–56

Anwar MM et al (2014) Economic deprivation of Indus River Delta, Sindh, Pakistan: causes and suggestions. Sci Int [Lahore] 26(2):885–890

AQUASTAT (2011) UN food and agriculture organization information system on water and agriculture, vol 37, Indus River Basin, Water Report. FAO, Rome. At: http://www.fao.org/nr/water/aquastat/basins/indus/index.stm

AQUASTAT (2013) UN food and agriculture organization information system on water and agriculture, vol 39, Afghanistan, water report. FAO, Rome. At: http://www.fao.org/nr/water/aquastat/countries_regions/afg/index.stm

AQUASTAT (2015) UN food and agriculture organization information system on water and agriculture. Main database. At: http://www.fao.org/nr/water/aquastat/data/query/index.html?lang=en

Asian Development Bank (2013) Thinking about water differently: managing the water-food-energy nexus. ADB, Mandaluyong City. At: http://www.adb.org/sites/default/files/publication/30409/thinking-about-water-differently.pdf

Bajracharya SR, Shrestha B (eds) (2011) The status of glaciers in the Hindu Kush-Himalayan region, ICIMOD, Kathmandu, November 2011. At: http://lib.icimod.org/record/9419/files/icimod-the_status_of_glaciers_in_the_hindu_kush-himalayan_region%5B1%5D.pdf

Bates, Bryson et al. (eds) (2008) Climate change and water, IPCC technical paper VI. Intergovernmental Panel on Climate Change, Geneva. At: http://ipcc.ch/pdf/technical-papers/climate-change-water-en.pdf

Beck MW et al (2012) Environmental and livelihood impacts of dams: common lessons across development gradients that challenge sustainability. Int J River Basin Manag 10(1):73–92. At: http://www.consbio.umn.edu/download/Beck_et_al2012.pdf

Benson D et al (2015) Water governance in comparative perspective: from IWRM to a 'Nexus' approach. Water Altern 8(1):756–773. At: http://www.water-alternatives.org/index.php/alldoc/articles/vol8/v8issue1/275-a8-1-8/file

Bolch T et al (2012) The state and fate of the Himalayan glaciers. Science 336(310):310–314

Briscoe J (2009) Water security: why it matters and what to do about it. Innovations 4(3):3–28. At: http://johnbriscoe.seas.harvard.edu/files/johnbriscoe/files/106._briscoe-_water_security-_lead_essay_in_innovations-_mit-_2009.pdf?m=1393430786

Chaturvedi AK (2013) Water: a source for future conflict. Vij Books, New Delhi

Chauhan BS et al (2014) Global warming and its possible impact on agriculture in India. Adv Agron 123(2):65–121

Cheema M, Pawar P (2015) Bridging the divide: transboundary science & policy interaction in the Indus basin. Stimson Center, Washington, DC. At: http://www.stimson.org/images/uploads/research-pdfs/EnvSecVFPaperMarch2015-FINAL.pdf

Christensen JH et al (2013) Climate phenomena and their relevance for future regional climate change. In: Stocker TF et al (eds) Climate change 2013: the physical science basis. Contribution

of working group I to the fifth assessment report of the intergovernmental panel on climate change. Cambridge University Press, Cambridge. At: http://www.ipcc.ch/report/ar5/wg1/
Church JA et al (2013) Sea level change. In: Stocker TF et al (eds) Climate change 2013: the physical science basis. Contribution of working group I to the fifth assessment report of the intergovernmental panel on climate change. Cambridge University Press, Cambridge. At: http://www.ipcc.ch/report/ar5/wg1/
Dankers R et al (2014) First look at changes in flood hazard in the inter-sectoral impact model intercomparison project ensemble. Proc Natl Acad Sci 111(9):3257–3261. At: http://www.pnas.org/content/111/9/3257.full.pdf
Department of Climate Change (2015) Enhanced actions on climate change: China's intended nationally determined contribution. National Development and Reform Commission, Beijing. At: http://www4.unfccc.int/submissions/INDC/Submission%20Pages/submissions.aspx
Dhillon BS et al (2010) National food security vis-à-vis sustainability of agriculture in high crop productivity regions. Curr Sci 98(1):33–36. At: http://re.indiaenvironmentportal.org.in/files/National%20food%20security.pdf
Dixit Y et al (2014) Abrupt weakening of the summer monsoon in northwest India ~4100 yr ago. Geology 42(4):339–342. At: http://geology.gsapubs.org/content/early/2014/02/24/G35236.1.full.pdf
Duflo E, Pande R (2007) Dams. Q J Econ 122(2):601–646
Ebrahim ZT (2015) Pakistan offers nothing to Paris climate summit. Thirdpole.net. 18 November 2015. At: http://www.thethirdpole.net/2015/11/18/pakistan-offers-nothing-to-paris-climate-summit/
Economy E (2011) China's growing water crisis. World Polit Rev. 9 Aug 2011. At: http://www.worldpoliticsreview.com/articles/9684/chinas-growing-water-crisis
Erenstein O (2009) Comparing water management in rice-wheat production systems in Haryana, India and Punjab, Pakistan. Agric Water Manag 96(12):1799–1806
Fischer EM, Knutti R (2015) Anthropogenic contribution to global occurrence of heavy-precipitation and high-temperature extremes. Nat Clim Chang 5(560):560–564
Fischer EM et al (2013) Robust spatially aggregated projections of climate extremes. Nat Clim Chang 3(1033):1033–1038. At: http://applcc.org/conservation-design/climate-context/uncertainty/documents/ClimateSciPDFs/fischerNCC2013.pdf/view
Food and Agricultural Organization of the United Nations (2014) The water-energy-food nexus: a new approach in support of food security and sustainable agriculture. FAO, Rome. At: http://www.fao.org/nr/water/docs/FAO_nexus_concept.pdf
Frenken K (ed) (2012) Irrigation in Southern and Eastern Asia in figures: AQUASTAT survey. FAO, Rome. At: http://www.fao.org/docrep/016/i2809e/i2809e.pdf
Giosan L et al (2012) Fluvial landscapes of the Harappan civilization. Proc Natl Acad Sci 109(26):E1688–E1694. At: http://www.pnas.org/content/109/26/E1688.full.pdf
Government of India (2015) India's intended nationally determined contribution: working toward climate justice. At: http://www4.unfccc.int/submissions/INDC/Submission%20Pages/submissions.aspx
Grey D, Sadoff C (2007) Sink or swim: water security for growth and development. Water Policy 9(6):545–571. At: http://cip.management.dal.ca/publications/Water%20security%20for%20growth%20and%20development.pdf
Haider M (2015) Ministry directed to finalise national water policy. The News. 1 September 2015. At: http://www.thenews.com.pk/Todays-News-13-39388-Ministry-directed-to-finalise-national-water-policy
Hansen J et al (2012) Perception of climate change. Proc Natl Acad Sci 109(37):14726–14727. At: http://www.pnas.org/content/109/37/E2415.full.pdf?with-ds=yes
Hao Z et al (2013) Changes in concurrent monthly precipitation and temperature extremes. Environ Res Lett 8(034014):1–7. At: http://iopscience.iop.org/article/10.1088/1748-9326/8/3/034014/pdf

Hartmann H, Buchanan H (2014) Trends in extreme precipitation events in the Indus river basin and flooding in Pakistan. Atmosphere-Ocean 52(1):77–91

Hartmann DL et al (2013) Observations: atmosphere and surface. In: Stocker TF et al (eds) Climate change 2013: the physical science basis. Contribution of working group I to the fifth assessment report of the intergovernmental panel on climate change. Cambridge University Press, Cambridge. At: http://www.ipcc.ch/report/ar5/wg1/

Hejazi MI et al (2014) Integrated assessment of global water scarcity over the 21st century under multiple climate change mitigation policies. Hydrol Earth Syst Sci 18(2859):2859–2883. At: http://www.hydrol-earth-syst-sci.net/18/2859/2014/hess-18-2859-2014.pdf

Hijioka Y et al (2014) Asia. In: Barros VR et al (eds) Climate change 2014: impacts, adaptation and vulnerability. Part B: regional aspects. Contribution of working group II to the fifth assessment report of the intergovernmental panel on climate change. Cambridge University Press, Cambridge. At: http://www.ipcc.ch/report/ar5/wg2/

Hornby C (2012) Solve water problems or forget growth, India Told. Reuters. 30 April 2012. At: http://www.reuters.com/article/us-india-water-idUSBRE83T0QV20120430

Immerzeel WW et al (2010) Climate change will affect the Asian water towers. Science 328(5984):1382–1385. At: http://www.ifu.ethz.ch/hydrologie/education/master_courses/glaciers/ImmerzeelScience

Imran A et al. (2014) An analytical study of the variations in the monsoon patterns over Pakistan. Pak J Meteorol 10:(20). At: http://www.researchgate.net/publication/269274468_An_Analytical_Study_of_the_Variations_in_the_Monsoon_Patterns_over_Pakistan_Imran_A_Q._Zaman

INBO (International Network of Basin Organizations) (2013) Fortaleza declaration, 9th world general assembly of the international network of basin organizations, Fortaleza, Brazil. 13–16 Aug 2013. At: http://www.inbo-news.org/IMG/pdf/EN_INBO_GA_FORTALEZA_2013_-_Final_DECLARATION-2.pdf

Indus Basin Working Group (2013) Connecting the drops: an Indus basin roadmap for cross-border water research, data sharing, and policy coordination. ORF/Stimson/SDPI, Washington, DC. At: http://www.stimson.org/images/uploads/research-pdfs/connecting_the_drops_stimson.pdf

International Centre for Integrated Mountain Development (2012) Status of glaciers in the Indus basin. ICIMOD, Kathmandu. At: http://lib.icimod.org/record/27040

Islamic Republic of Afghanistan (2015) Intended nationally determined contribution: submission to the United Nations framework convention on climate change. 21 September 2015. At: http://www4.unfccc.int/submissions/INDC/Submission%20Pages/submissions.aspx

Ives JD et al (2010) Formation of glacial lakes in the Hindu Kush-Himalayas and GLOF risk assessment. ICIMOD, Kathmandu. At: http://www.unisdr.org/files/14048_ICIMODGLOF.pdf

Jiménez Cisneros B et al (2014) Freshwater resources. In: Fields CB et al (eds) Climate change 2014: impacts, adaptation and vulnerability. Part a: global and sectoral aspects. Contribution of working group II to the fifth assessment report of the intergovernmental panel on climate change. Cambridge University Press, Cambridge. At: http://www.ipcc.ch/report/ar5/wg2/

Kääb A et al (2012) Contrasting patterns of early twenty-first-century glacier mass change in the Himalayas. Nature 488(495):495–498. At: http://etienne.berthier.free.fr/download/Kaab_et_al_Nature_2012.pdf

Kääb A et al (2015) Brief communication: contending estimates of 2003–2008 glacier mass balance over the Pamir-Karakoram-Himalaya. Cryosphere 9(557):557–564. At: http://www.the-cryosphere.net/9/557/2015/tc-9-557-2015.html

Karki MB et al (2011) Enhancing knowledge management and adaptation capacity for integrated management of water resources in the Indus river basin. Mt Res Dev 31(3):242–251. At: http://www.bioone.org/doi/pdf/10.1659/MRD-JOURNAL-D-11-00017.1

Kravtsova VI et al (2009) Variations of the hydrological regime, morphological structure, and landscapes of the Indus river delta (Pakistan) under the effect of large-scale water management measures. Water Res 36(4):365–379

Kugelman M, Hathaway RM (eds) (2009) Running on empty: Pakistan's water crisis. Woodrow Wilson International Center for Scholars, Washington, DC. At: https://www.wilsoncenter.org/sites/default/files/ASIA_090422_Running%20on%20Empty_web.pdf
Kulkarni AV, Karyakarte Y (2014) Observed changes in Himalayan glaciers. Curr Sci 106(2):237–244. At: http://www.dccc.iisc.ernet.in/observed%20change.pdf
Kundzewicz ZW et al (2014) Flood risk and climate change: global and regional perspectives. Hydrol Sci J 59(1):1–28. At: https://floodobservatory.colorado.edu/Publications/Kundzewicz.pdf
Laghari JR (2013) Melting glaciers bring energy uncertainty. Nature 502(617):617–618. At: http://www.nature.com/polopoly_fs/1.14031!/menu/main/topColumns/topLeftColumn/pdf/502617a.pdf
Laghari AN et al (2012) The Indus basin in the framework of current and future water resources management. Hydrol Earth Syst Sci 16(4):1063–1083. At: http://www.hydrol-earth-syst-sci.net/16/1063/2012/hess-16-1063-2012.pdf
Mack TJ et al (2010) Conceptual model of water resources in the Kabul basin, Afghanistan, scientific investigations report 2009–5262. U.S. Geological Survey, Reston. At: http://pubs.usgs.gov/sir/2009/5262/
Maini TS (2012) The two Punjabs: one step more towards closer cooperation? East Asia Forum. 14 Nov 2012. At: http://www.eastasiaforum.org/2012/11/14/the-two-punjabs-one-step-more-toward-closer-cooperation/
Masood S (2015) Starved for energy, Pakistan braces for a water crisis. N Y Times. 12 February 2015. At: http://www.nytimes.com/2015/02/13/world/asia/pakistan-braces-for-major-water--shortages.html?_r=0
Meinshausen M et al (2011) The RCP greenhouse gas concentrations and their extensions from 1765–2300. Clim Chang 109 (1 and 2):213–241. At: http://link.springer.com/article/10.1007/s10584-011-0156-z
Milly PCD et al (2008) Climate science – stationarity is dead: whither water management? Science 319(573):573–574. At: http://aquadoc.typepad.com/waterwired/files/milly_et_al.pdf
Ministry of Climate Change (2012) National climate change policy. Government of Pakistan, Islamabad. At: http://www.mocc.gov.pk/
Ministry of Climate Change (2015a) Pakistan Intended Nationally Determined Contributions (INDCs). Zero Draft n.d. At: http://www.mocc.gov.pk/
Ministry of Climate Change (2015b) Pakistan Intended Nationally Determined Contributions (INDCs). n.d. At: http://www4.unfccc.int/submissions/INDC/Submission%20Pages/submissions.aspx
Ministry of Environment and Forests (2010) Climate change and India: a 4x4 assessment – a sectoral and regional analysis for 2030s. Government of India, New Delhi. At: http://www.moef.nic.in/downloads/public-information/fin-rpt-incca.pdf
Ministry of Environment and Forests (2012) India: second national communication to the United Nations framework convention on climate change. Government of India, New Delhi. At: http://unfccc.int/national_reports/non-annex_i_natcom/items/2979.php
Ministry of Water Resources (2012) National water policy. Government of India, New Delhi. At: http://wrmin.nic.in/writereaddata/NationalWaterPolicy/NWP2012Eng6495132651.pdf
Ministry of Water Resources (2015) Status of Memorandum of Understanding (MoU) signed with foreign countries in the water sector, New Delhi, Government of India. At: http://wrmin.nic.in/writereaddata/MOU_ForeignCountries_Status.pdf
Molle F et al (2010) River basin closure: processes, implications and responses. Agric Water Manag 97(4):569–577
Mukhopadhyay B (2012) Detection of dual effects of degradation of perennial snow and ice covers on the hydrologic regime of a Himalayan River basin by stream water availability modeling. J Hydrol 412–413(14). At: http://www.researchgate.net/publication/251473202_Detection_of_dual_effects_of_degradation_of_perennial_snow_and_ice_covers_on_the_hydrologic_regime_of_a_Himalayan_river_basin_by_stream_water_availability_modeling

Mukhopadhyay B, Khan A (2015) A reevaluation of the snowmelt and glacial melt in river flows within the upper Indus basin and its significance in a changing climate. J Hydrol 527(119)
National Development and Reform Commission (2012) Second national communication on climate change of the People's Republic of China. 8 Nov 2012. At: http://unfccc.int/national_reports/non-annex_i_natcom/items/2979.php
National Development and Reform Commission (2014) China's policies and actions on climate change. National Development and Reform Commission, Beijing. At: http://en.ccchina.gov.cn/list.aspx?clmId=102
National Environmental Protection Agency (2009) Afghanistan: National Capacity Needs Self-Assessment for Global Environmental Management (NCSA) and National Adaptation Programme of Action for climate change (NAPA) – final joint report. UNEP/NEPA/GEF, Nairobi. At: http://unfccc.int/adaptation/workstreams/national_adaptation_programmes_of_action/items/4585.php
National Environmental Protection Agency (2012) Afghanistan initial national communication to the United Nations framework convention on climate change. NEPA, Kabul. At: http://unfccc.int/essential_background/library/items/3599.php?rec=j&priref=7713#beg
National Research Council (2012) Himalayan glaciers: climate change, water resources, and water security. National Academies Press, Washington, DC. At: http://dels.nas.edu/Report/Himalayan-Glaciers-Climate-Change-Water-Resources/13449
Nepal S, Shrestha AB (2015) Impact of climate change on the hydrological regime of the Indus, Ganges and Brahmaputra river basins: a review of the literature. Int J Water Resour Dev 31(2):201–218. At: http://www.tandfonline.com/doi/abs/10.1080/07900627.2015.1030494
OECD (2015) OECD principles on water governance. Organisation for Economic Cooperation and Development, Paris. At: http://www.oecd.org/gov/regional-policy/OECD-Principles-on-Water-Governance-brochure.pdf
Pal I, Al-Tabaa A (2010) Regional changes in extreme monsoon rainfall deficit and excess in India. Dyn Oceans Atmos 49(2 and 3):206–214
Palmer MA et al (2008) Climate change and the World's river basins: anticipating management options. Front Ecol Environ 6(2):81–89
Perveen S et al. (2012) India's deepening water crisis? Water risks for Indian industries, federation of Indian chambers of commerce and industry/Columbia University Water Center. Feb 2012. At: http://water.columbia.edu/files/2012/06/FICCI_CWC_IndiaWaterCrisisPaper.pdf
Pew Research Center (2015) Global concern about climate change, broad support for limiting emissions. Nov 2015. At: http://www.pewglobal.org/files/2015/11/Pew-Research-Center-Climate-Change-Report-FINAL-November-5-2015.pdf
Planning Commission of Pakistan (2011) Pakistan framework for economic growth. Government of Pakistan, Islamabad. At: http://www.pc.gov.pk/hot%20links/growth_document_english_version.pdf
Praduman K et al (2014) Projected effect of droughts on supply, demand, and prices of crops in India. Econ Polit Wkly 49(52):54–63
Price G et al (2014) Attitudes to water in South Asia. Chatham House, London. At: https://www.chathamhouse.org/sites/files/chathamhouse/field/field_document/20140627WaterSouthAsia.pdf
Prime Minister's Council on Climate Change (2008) National action plan on climate change. Government of India, New Delhi. At: http://www.moef.nic.in/sites/default/files/Pg01-52_2.pdf
Qiu J (2012) Flood of protests hits Indian dams. Nature 292(15). At: http://www.indiaenvironmentportal.org.in/files/file/Indian%20dams.pdf
Qureshi AS (2011) Water management in the Indus basin in Pakistan: challenges and opportunities. Mt Res Dev 31(3):252–260. At: http://www.bioone.org/doi/pdf/10.1659/MRD-JOURNAL-D-11-00019.1
Radic V et al (2014) Regional and global projections of twenty-first century glacier mass changes in response to climate scenarios from global climate models. Clim Dyn 42(1):37–58. At: http://www.eos.ubc.ca/~vradic/Radic_et_al2013_ClimDyn.pdf

Rajbhandari R et al (2015) Projected changes in climate over the Indus river basin using a high resolution regional climate model (PRECIS). Clim Dyn 44(1):339–357. At: http://link.springer.com/article/10.1007/s00382-014-2183-8/fulltext.html

Rasul G et al. (2012) Vulnerability of the Indus delta to climate change in Pakistan. Pak J Meteorol 8(16):89–106. At: http://www.pmd.gov.pk/rnd/rnd_files/vol8_Issue16/8_Vulnerability%20of%20the%20Indus%20Delta%20to%20Climate%20Change%20in%20Pakistan.pdf

Reynolds JM (2014) Assessing glacial hazards for hydro development in the Himalayas, Hindu Kush and Karakoram. Hydropower Dams 2:60–65

Rhein M et al (2013) Observations: ocean. In: Stocker TF et al (eds) Climate change 2013: the physical science basis. Contribution of working group I to the fifth assessment report of the intergovernmental panel on climate change. Cambridge University Press, Cambridge. At http://www.ipcc.ch/report/ar5/wg1/

Richter BD et al (2010) Lost in development's shadow: the downstream consequences of dams. Water Altern 32(2):14–42. At: http://www.water-alternatives.org/index.php/volume3/v3issue2/80-a3-2-3/file

Rieu-Clarke A et al (2015) Transboundary water governance and climate change adaptation: international law, policy guidelines, and best practice application. UNESCO, Paris. At: http://unesdoc.unesco.org/images/0023/002356/235678e.pdf

Ringler C et al (2009) Water supply and food security: alternative scenarios for the Indian Indo-Gangetic river basin. Int J River Basin Manag 7(2):167–173. At: http://www.tandfonline.com/doi/pdf/10.1080/15715124.2009.9635379

Roxy MK et al (2015) Drying of Indian subcontinent by rapid Indian ocean warming and a weakening land-sea thermal gradient. Nat Commun 6(7423):7423

Sadoff C, Muller M (2009) Water management, water security and climate adaptation: early impacts and essential responses, vol 14, TEC Background Papers. Global Water Partnership, Stockholm. At: https://dlc.dlib.indiana.edu/dlc/bitstream/handle/10535/4999/GWP_TEC_14_FINAL.pdf?sequence=1

Savoskul OS, Smakhtin V (2013) Glacier systems and seasonal snow cover in six major Asian river basins: hydrological role under changing climate, vol 150, IWMI Research Report. International Water Management Institute, Colombo. At: http://www.iwmi.cgiar.org/Publications/IWMI_Research_Reports/PDF/PUB150/RR150.pdf

Schoeman J et al (2014) A new paradigm for water: a comparative review of integrated, adaptive, and ecosystem-based water management in the anthropocene. Int J Water Resour Dev 30(3):377–390

Sharma BR et al (2008) Indo-Gangetic river basins: summary situation analysis, international water management institute, New Delhi office. Jul 2008. At: http://cpwfbfp.pbworks.com/f/IGB_situation_analysis.PDF

Sheikh MM et al (2009) Climate profile and past climate changes in Pakistan, Research Report GCISC-RR-01. Global Change Impact Studies Centre, Islamabad

Shrestha MS et al (2015) Establishment of a regional flood information system in the Hindu Kush Himalayas: challenges and opportunities. Int J Water Resour Dev 31(2):238–252. At: http://www.tandfonline.com/doi/pdf/10.1080/07900627.2015.1023891

Shroder JF (2014) Natural resources in Afghanistan: geographic and geologic perspectives on centuries of conflict. Elsevier, San Diego

Siddiqi A, Wescoat JL (2013) Energy use in large-scale irrigated agriculture in the Punjab province of Pakistan. Water Int 38(5):571–586. At: http://www.tandfonline.com/doi/pdf/10.1080/02508060.2013.828671

Singh K (2008) City-to-city cooperation for sustainable urban development. Man & Dev 30(4):45–64. At: http://www.indiaenvironmentportal.org.in/files/Man%20Development%203.pdf

Singh SP et al (2011) Climate change in the Hindu Kush-Himalayas: the state of current knowledge. ICIMOD, Kathmandu. At: http://lib.icimod.org/record/9417/files/icimod-climate_change_in_the_hindu_kush-himalayas.pdf

Smakhtin V (2008) Basin closure and environmental flow requirements. Int J Water Resour Dev 24(2):227–233
Smakhtin V et al (2004) A pilot global assessment of environmental water requirements and scarcity. Water Int 29(3):307–317. At: https://www.uni-frankfurt.de/45217759/Smakhtin_revenga_doell_WaterInternational2004_EWR.pdf
Stancati M (2012) India's water waste could hurt growth. Wall Str J. 11 Apr 2012. At: http://blogs.wsj.com/indiarealtime/2012/04/11/indias-water-waste-could-hurt-growth/
Syvitski JPM, Brakenridge GR (2013) Causation and avoidance of catastrophic flooding along the Indus River, Pakistan. GSA Today 23(1):4–10. At: https://floodobservatory.colorado.edu/Publications/gsatv23n01_13-sc%5B1%5D.pdf
Syvitski JPM et al (2013) Anthropocene metamorphosis of the Indus delta and lower floodplain. Anthropocene 3:24–35
UNECE (2015) Water and climate change adaptation in transboundary basins: lessons learned and good practices. United Nations Economic Commission for Europe, Geneva. At: https://www.unece.org/fileadmin/DAM/env/water/publications/WAT_Good_practices/ece.mp.wat.45.pdf
Varis O et al (2012) Ten major rivers in monsoon Asia-Pacific: an assessment of vulnerability. Appl Geogr 32(2):441–454
Vaughan DG et al (2013) Observations: cryosphere. In: Stocker TF et al (eds) Climate change 2013: the physical science basis. Contribution of working group I to the fifth assessment report of the intergovernmental panel on climate change. Cambridge University Press, Cambridge. At: http://www.ipcc.ch/report/ar5/wg1/
Vick MJ (2014) Steps towards an Afghanistan-Pakistan water sharing treaty. Int J Water Resour Dev 30(2):224–229. At: http://www.tandfonline.com/doi/pdf/10.1080/07900627.2014.886471
Wada Y, Bierkens MFP (2014) Sustainability of global water use: past reconstruction and future projections. Environ Res Lett 9(104003):104003. At: http://iopscience.iop.org/article/10.1088/1748-9326/9/10/104003/pdf
Walker B et al (2014) China's water security crisis, Europe China research and advice network. Feb 2014
Warren Z, Hopkins N (eds) (2015) Afghanistan in 2015: a survey of the Afghan people. The Asia Foundation, San Francisco. At: http://asiafoundation.org/publications/pdf/1558
World Bank (2015) World development indicators. At: http://data.worldbank.org/indicator
WWAP (World Water Assessment Programme) (2012) The United Nations world water assessment report 4: managing water under uncertainty and risk. UNESCO, Paris. At: http://www.unesco.org/new/en/natural-sciences/environment/water/wwap/wwdr/wwdr4-2012/
Xinhua (2015) China's silk road fund makes first investment in Pakistan's hydropower project, Xinhua. 21 April 2015. At: http://news.xinhuanet.com/english/2015-04/21/c_134167533.htm
Yu W et al (2013) The Indus basin of Pakistan: the impacts of climate risks on water and agriculture. World Bank, Washington, DC. At: http://www-wds.worldbank.org/external/default/WDSContentServer/WDSP/IB/2013/05/06/000445729_20130506101316/Rendered/PDF/770220PUB0REPL0te040180130EPI019874.pdf
Zemp M et al (2015) Historically unprecedented glacier decline in the early 21st century. J Glaciol 61(228):745–762. At: http://www.geo.uzh.ch/~mzemp/Docs/Zemp_etal_JoG_2015.pdf
Zia A, Wagner CH (2015) Mainstreaming early warning systems in development and planning processes: multilevel implementation of Sendai framework in Indus and Sahel. Int J Dis Risk Sci 6(2):189–199. At: http://link.springer.com/article/10.1007%2Fs13753-015-0048-3
Zuo Q et al (2014) China pursues a strict water resources management system. Environ Earth Sci 72(6):2219–2222

Chapter 7
Transboundary Data Sharing and Resilience Scenarios: Harnessing the Role of Regional Organizations for Environmental Security

Saleem H. Ali and Asim Zia

Abstract Effective regional planning in the Indus basin necessitates greater transboundary data sharing on environmental indicators. Reliable hydrological data across political borders is essential to ensure more accurate and effective risk management mechanisms. Given the high vulnerability of this region to climate change, data sharing through existing regional cooperation organizations such as South Asian Association for Regional Cooperation (SAARC) and International Centre for Integrated Mountain Development (ICIMOD) must be further encouraged. There is potential to also learn from the limitations of the data sharing mechanisms under the Indus Waters Treaty. Recent natural disasters have highlighted the need for urgent action on this matter, and thus the vision of "Industan" is not merely a peacebuilding ideal but a pragmatic risk management strategy, particularly for India and Pakistan.

Keywords Environmental security • Science diplomacy • Regional cooperation • Climate change

7.1 Introduction

South Asia is home to a quarter of the world's population, inhabiting some of the planet's most diverse ecological systems, from the highest mountain range (the Himalayas) to the largest riparian delta system (the Ganges-Brahmaputra). By 2050, South Asia's population will exceed 2.2 billion, with an estimated 600 million people living on less than $1.25 a day. About 70 % of South Asians live in rural areas,

S.H. Ali (✉)
University of Queensland, Brisbane, Australia
e-mail: s.ali3@uq.edu.au

A. Zia
University of Vermont, Burlington, Vermont, USA
e-mail: Asim.Zia@uvm.edu

Z. Adeel, R.G. Wirsing (eds.), *Imagining Industan*, Water Security in a New World, DOI 10.1007/978-3-319-32845-4_7

representing 75 % of those at the lowest income levels.[1] Given this population distribution and the relatively rapid change in the environmental profile of South Asia, the human vulnerability to even minor environmental stresses and consequential conflict and civil strife is very high. The purpose of this study is to explore ways by which the planning for food, water, and energy nexus in South Asia might be shifted from conflict to cooperation using ecological factors as a binding mechanism. Environmental and science diplomacy can play a catalytic role in transforming this nexus, enabling nation-states to rise above their sovereignty concerns and reap the benefits of cooperation in the form of food, water, and energy security in the face of daunting challenges posed by population, climate change, and other stresses. The environment provides a superordinate goal whose parameters of impact are beyond the control of antagonists. Scientific cooperation can further the chance of both sides, understanding and mitigating the negative impact of natural or anthropogenic developments.

This study was carried out through a detailed and systematic review of national statements, conflict narratives in news stories and speeches of leaders, statements by regional organizations, and select interviews with stakeholders to gain clarification and context on particular events. Earlier research by the authors on data sharing was also incorporated in the policy prescriptions offered.

7.1.1 Greening Existing Regional Organizations: Beyond Historical Inertia

Regional cooperation in South Asia is a relatively recent phenomenon. While South and East Asia were busy working on visa-free trade zones through organizations such as the Association of Southeast Asian Nations (ASEAN), much of South Asia was embroiled in conflicts. As with other parts of the world, the Cold War created a polarization that prevented regional cooperation. India, the dominant power, was focused instead on developing the Non-Aligned Movement as an antidote to Cold War allegiances. It was not until 1980 that the idea of establishing a separate organization focused on South Asian cooperation was moved forward. The preconditions for establishing the South Asian Association for Regional Cooperation (SAARC) were quite confining in terms of what could be achieved but similar to those of several other organizations for regional cooperation conceived at the time. Five key principles define all SAARC activities:

Respect for sovereignty, territorial integrity, political equality, and independence of all member states
Noninterference in internal matters
Cooperation for mutual benefit

[1] World Bank estimates derived from South Asia portal on World Bank Web site – www.worldbank.org

All decisions to be taken unanimously and needing a quorum of all eight members
All bilateral issues to be kept aside and only multilateral (involving many countries) issues to be discussed without being prejudiced by bilateral issues

SAARC was formally established as a permanent organization in 1985, with a secretariat hosted in Kathmandu, Nepal. Its seven original members—Bangladesh, Bhutan, India, Maldives, Nepal, Pakistan, and Sri Lanka—agreed to admit Afghanistan as an eighth member in 2007.[2] The addition was particularly significant because SAARC could thereby act as a forum for India and Pakistan to negotiate their strategic influence over Afghanistan's development path. In Pakistan, there has been recurring suspicion about ulterior motives for India's high level of development aid to Afghanistan. Allowing for a transparent exchange on regional development investment in Afghanistan could be an effective means of assuaging some of this mistrust. At the same time, there has been movement by Pakistan and India to establish their own spheres of influence, heading west and east, respectively, for regional partners.

The Bay of Bengal Initiative for Multi-Sectoral Technical and Economic Cooperation (BIMSTEC) was formed partly as a response to the perceived inefficacy of SAARC due to repeated deadlocks between India and Pakistan. The BIMSTEC agreement also includes Myanmar and Thailand, with which India has strong trade ties. Energy and some environmental areas of cooperation, such as fisheries, are proposed subjects of activity, in addition to a broader trade and infrastructure cooperation agenda. However, there is still modest progress on ecological cooperation, although the Asian Development Bank has engaged with BIMSTEC to support infrastructure linkages, particularly in the transport sector, whereby environmental planning criteria may be more directly incorporated.

On the western frontier, the Economic Cooperation Organization (ECO) has a specific directorate pertaining to minerals, energy, and environment. This organization was established in the same year as SAARC (1985) by Iran, Pakistan, and Turkey with the goals of promoting economic, cultural, and technical cooperation. With the dissolution of the Soviet Union, several Central Asian states joined ECO, and the organization now comprises ten member states. All members are Muslim-majority states and six of the ten are landlocked. ECO is a forum for Afghanistan and Pakistan to have more direct communication facilitated by Central Asian partners and is considered a counterweight to the dominance of India in communications within SAARC. However, Iran's involvement in the organization and the fact that its secretariat is based in Tehran make it difficult to gain much interest from international donors despite some important proposed initiatives related to regional ecotourism and energy infrastructure collaboration (Pomfret 1997).

One possible connection that could be made between ECO and SAARC was suggested by then Pakistani prime minister Yousuf Raza Gillani at the SAARC summit

[2] Recognizing the importance of development donors and broader strategic interests, SAARC has also granted observer status to Australia, China, the European Union, Japan, Iran, Mauritius, Myanmar, South Korea, and the United States (Ahmed 2013).

in 2011 and involved the road link between Tajikistan and Pakistan to connect South Asian markets to SAARC countries.[3] This broader vision would also tie in with the "New Silk Road" initiative that US Secretary of State Hillary Clinton announced at a regional meeting in Chennai in 2011. The goals of the Silk Road vision were elaborated by the State Department official Geoffrey Pyatt in an official policy speech at a conference in Tokyo in June 2012, in which he noted that in addition to the expansion of merchandise flow, the vision involves cooperation "through energy, water, transport, and infrastructure—which includes roads, bridges, electrical transmission grids, railways, and pipelines—*to connect* goods, services, and people" (Pyatt 2012).

However, to make such a Silk Road work better and with far greater consequence, undertaking these initiatives will inevitably need the assistance of China, which has established its own regional grouping that overlaps with the Central Asian members of ECO. The Shanghai Cooperation Organization (SCO), established as a means of strengthening China's partnership with Russia in the region, also has environmental and energy cooperation dimensions that could have a bearing on South Asian regional cooperation.[4] Among SAARC countries, India, Pakistan, and Afghanistan remain as "observers" within the SCO. Pakistan and India have actively lobbied for full SCO membership, and it was announced in July 2015 that both states will be granted full member status in 2016 (Reuters 2015). SCO has the potential of becoming a much more consequential partner in areas of energy and transport cooperation as the northern Asian states seek access to the lucrative high demographic growth markets of South Asia.

Despite the growth of these regional organizations, the potential for SAARC to play a role in multilateral ecological cooperation remains strongest. Within the ecological arena, SAARC has a program of work on environment and energy that includes the aims of establishing a specific convention on environmental cooperation, which was reaffirmed in the Thimphu Declaration on Climate Change (2010). Among the lesser-known accomplishments of SAARC is the establishment of the South Asian University in New Delhi,[5] where students from all member countries study together under one institutional umbrella. The university held its first classes in 2010, just 5 years after the idea was introduced at the SAARC summit in Dhaka, initially offering master's degree programs in computer applications and development economics. At the 2011 SAARC summit, Indian prime minister Manmohan Singh announced that India would increase the number of SAARC Silver Jubilee scholarships at the university from 50 to 100 (75 at the master's level and 25 at the doctoral level).

The persistent acrimony and nuclear rivalry between India and Pakistan have often hampered substantive progress on regional cooperation. Yet SAARC is evolving into a forum that links civil society and governments in the region through common denominators such as education, the environment, and human rights. At the

[3] "Pakistan PM to take-up Pak-Tajik road project." *The Daily Times* (Pakistan), October 10, 2011.

[4] For a review of SCO in the context of regional environmental cooperation, see Wang, 2011

[5] South Asian University, New Delhi, Web site accessed September 1, 2015: http://www.sau.int/

2011 summit, "Peoples' SAARC,"[6] a parallel initiative to the official SAARC that was established in 1996 as a means of providing policy evaluation to local governments, provided a "memorandum" with detailed practical "demands" concerning the rights of fishermen in regional waters, migratory populations, and communities affected by climate changes and disasters.

In his formal remarks at the SAARC summit in 2011, the Indian prime minister also stated unequivocally that India has "a special responsibility that flows from the geography of our region and the state of our economy and market."[7] Environmental cooperation was highlighted specifically in the context of the India Endowment for Climate Change, which will provide ten scholarships per year to citizens of SAARC member states for postgraduate and doctoral studies in forestry courses at the Forest Research Institute in Dehradun, India. The recognition that mountain ecology can be a binding educational mechanism is reflected by the choice of venue for this program.

7.2 Data Sharing as a Means of Science Diplomacy

Going back to 1991, when the South Asian Association for Regional Cooperation (SAARC) launched a *Regional Study on the Causes and Consequences of Natural Disasters and the Protection and Preservation of the Environment*, there has been a recognition that regional environmental cooperation is linked to the Himalayas—the world's highest mountain range, which defines the region's geography. As noted in the report at the time, "Pakistan, India, Nepal, and Bhutan share amongst themselves the vast Himalayan mountain range and Bangladesh's ecological situation is such that it is greatly influenced by ecological changes in the Himalaya. These mountains are today one of the most densely populated in the world and face severe human-made environmental problems together with natural hazards inherent in local ecological conditions" (SAARC 1992, p. 206). From an economic development perspective, the importance of linking climate change and leveraging the commonality of the Himalayas has also been noted by the Asian Development Bank in a commissioned paper on the *Political Economy of Regional Cooperation in South Asia* (Desai 2010).

Mountains have been considered natural borders and zones of separation. They often form physical barriers between human settlements and have thus defined cultural identities and formed political borders. Yet, environmental factors have led erstwhile adversaries across the cultural and political divide to consider ways of cooperating around mountain systems. This is due to the seminal role mountains

---

[6] Peoples' SAARC Web site, accessed September 2, 2012; http://www.peoplesaarc.org

[7] Indian prime minister Manmohan Singh's statement on NDTV Web site, accessed September 2, 2012; http://www.ndtv.com/india-news/prime-minister-manmohan-singh-speaks-at-the-saarc-summit-full-text-570066

play in providing resources for human survival—most notably their role in regulating climatic conditions through altitudinal variation.

Atmospheric water resources are brought to land most often through mountain systems. The hydrological fortune of countries and communities is thus often defined by which side of a mountain range they lie on and which crops they can cultivate. The "rain shadow," which has often determined the sparseness of populations, particularly in central and southern Asia, is determined almost entirely by the geography of mountain systems. The Himalayas traverse the most populated parts of the world, encompassing almost half of the world's population. Yet these mountains also define some of the most intractable territorial disputes in the region, particularly between India and Pakistan but also between India and China.

The most direct willingness to support the engagement of Indian and Pakistani scientists on cooperative research has come from the United States. With the support of the US National Science Foundation, International Centre for Integrated Mountain Development (ICIMOD) hosted a workshop for Indian and Pakistani glaciologists in 2007 to foster cooperative scientific research. This effort also tied in with the recurring calls for using environmental peace-building in resolving the Siachen dispute (Ali 2012; Bagla 2006a, b). However, no progress has occurred thus far on this front despite renewed impetus for such an undertaking after the tragic avalanche that killed more than 100 Pakistani soldiers and civilians at Siachen in April 2012.

Visa access for Indian and Pakistani scientists is far more difficult to obtain in comparison with art and cultural exchanges. There is still a perceived threat from scientific cooperation due to concerns about scientists getting access to sensitive security information. Even though environmental scientists focus on planetary processes rather than particular mechanical details of military devices, they are perceived to have the potential skills to transfer such sensitive information. Scientists have repeatedly been prevented access by both countries for collaborative meetings in this regard despite various cultural exchange visa programs. Perhaps the closest that Indian and Pakistani scientists have come to a concerted collaboration has been through the South Asian Network for Development and Environmental Economics (SANDEE), which was established in 1999 under the auspices of the International Union for Conservation of Nature (IUCN) and subsequently hosted at the ICIMOD secretariat. The mission of SANDEE is to "use economic tools and analyses to address South Asia's environmental challenges. It is based on the premise that solutions to economic development concerns and environmental problems are integrally linked."[8] The network has been well-resourced through grants from the World Bank, the Canadian government's International Development Research Centre (IDRC), the Norwegian Agency for Development Cooperation (NORAD), and the Swedish International Development Cooperation Agency (SIDA). A regular collaborative research grant program and numerous training programs across the region have been sponsored.

[8] South Asian Network for Development Economics and the Environment, Kathmandu, Nepal: http://www.sandeeonline.org/, retrieved August 10, 2015.

SANDEE's competitive grants program in 2011 generated 15 new projects out of a possible 99 pre-proposals. The grants spanned a variety of issues—biodiversity conservation, pollution effects, sustainable agriculture, water conservation, and policy analyses—with a third of them focused on climate change. SANDEE launched three cross-country studies on climate and migration in Bangladesh, India, and Pakistan in which it sought to understand the extent to which the movement of people is induced by the effects of weather on agriculture. This work is aimed at complementing ongoing studies about the burning of agricultural field residue in Bangladesh, India, Nepal, and Pakistan.

Collaborative research among scholars across national divides still remains elusive. Leadership from international organizations will be needed in directly focusing on Indo-Pak cooperative research, rather than simply couching the matter in regional cooperation terms, to have tangible impact on using such efforts as confidence-building mechanisms or citizen diplomacy.

7.3 Vulnerability in the Indus Basin

Global warming poses increasing risk to the millions of vulnerable people in the Indus basin, as it is expected that global warming-induced natural hazards, such as floods, droughts, heat waves, and cyclones, will increase in their frequency and intensity as the twenty-first century unfolds. Just alone the very likely meltdown of the great Himalayan ice mass will cause a big disturbance in the hydrometeorological cycle of the Indus system, the lifeline for millions of people's food security, hydropower, and fresh water availability (Immerzeel et al. 2010). Climate change-induced higher frequencies and intensities of both primary and consequent extreme events such as flooding-induced landslides will be compounded by already stressed political, social, and economic conditions across the four countries that comprise the Indus basin (Zia and Wagner-Hammond 2015). The eco-hydrological boundaries of the upper Indus basin in Himalayas cut across administrative boundaries of China, India, Pakistan, and Afghanistan. However, the global climatic and environmental changes cut across these administrative boundaries; and detailed impact and vulnerability analyses of global environmental change-induced risks at the local planning and policy scale require much more improved regional scale integrated assessment models. Aggregate modeling results at the global scale are currently available in the Intergovernmental Panel on Climate Change's Fourth Assessment Report (IPCC FAR). The global scale coarser grid results are typically ascertained at 1° latitude × 1° longitude spatial scale from global climate models (GCMs). The coarser grid scenarios project a grim picture for the food-energy-water nexus in the Indus basin under representative concentration pathways (RCP) 8.5 W/m^2 of global warming radiance set of scenario ensembles. Chapter 24 in IPCC FAR working group II, which focuses on impacts, adaptation, and vulnerability in Asia, summarizes the extent of vulnerability in the food system under different climate change scenarios ascertained from GCM runs. In this chapter, Hijioka et al. (2014:1330) conclude that "In the Indo-Gangetic Plains of South Asia there could be a decrease of about 50 % in the most favorable and high-yielding wheat area as a result of heat stress at 2 times CO_2."

Our previous research on the Indus basin, discussed in Zia and Glantz (2012), Zia (2013), and Zia and Wagner-Hammond (2015), suggests that uncertainties about monsoon variability and glacial melt timing pose enormous planning and policy challenges for the Indus basin. The Indus basin is a breadbasket for millions of people across the subcontinent, a major source of hydropower and freshwater every year, yet glacial melt and monsoon variability threatens the future of food-energy-water availability in its catchment regions.

While the coarser resolution global scale climate impact assessments suggest the likelihood of radical shifts in eco-hydrological regimes that initiate a cascade of downstream impacts on food-energy-water systems of the Indus basin, more accurate projections at finer space-time resolution are direly needed for proactive policy and planning interventions to mitigate the risk from climate change-induced extreme events. This would, however, require development of calibrated eco-hydrological models at the basin scale, and their integration with socioeconomic models, to estimate the risk posed by climate change-induced extreme events and identify leverage points for policy makers and planners to mitigate the risk with targeted interventions. Such regional scale models need to be calibrated with fine-scaled sub-watershed level precipitation and discharge data. Given the transboundary nature of the upper Indus basin, as well as the importance of including accurate glacial melt data in the regional scale models, it is critical that the four countries in the upper Indus share and exchange the underlying eco-hydrological data.

7.4 Data Gaps and Related Uncertainties

In the current geopolitical situation, there is a trust deficit in data sharing and data collection coordination across the four countries of the Indus basin. Both India and China have had regular sets of skirmishes in the Ladakh region, which, coupled with the Tibet problem, has led to military and diplomatic skirmishes between India and China. The challenges between India and Pakistan that range from Kashmir to Siachen, allegations of terrorism, and the recent Indo-Pak struggle for ascendant influence in Afghanistan must also be taken into account for explaining the prevalence of data sharing trust deficit. The capacity of regional organizations, such as SAARC and ICIMOD, in promoting data sharing across the regions needs to be enhanced and beefed up. Under the Indus Waters Treaty (IWT), aggregate flow statistics of the six Indus tributaries are reported by Indian and Pakistan agencies to the World Bank. These aggregate statistics can be used for the validation of Indus basin-wide hydrological models in historical context; however, this aggregate data does not provide adequate information to calibrate the internal components of the hydrological models that require spatially distributed hydrological and land-use/land-cover change data at finer resolutions (i.e., less than 1 km × km spatial resolutions). In a recent study of the upper Indus basin, Ali et al. (2015: 18) found that the downscaled climatic and hydrological models – the CCAM and RegCM model – "underestimate the temperature, whereas they overestimate the precipitation over UIB (Upper Indus Basin)." While Ali et al. used statistical bias correction methods, such methods add to the uncertainty of projections under different climate change scenarios. Higher uncertainty of these estimates makes it difficult to draw meaningful inferences for water-energy-food nexus planning at the Indus basin level. A better, durable approach to resolve this problem, when expanding such models from upper Indus basin to the entire basin, would be to pool decentralized spatial databases across the four riparian states. While satellite and remote sensing databases can provide increasingly useful information about land-use/land-cover change as well as changes in the upper atmosphere, such databases cannot provide information about the hydrological (e.g., stream flow) and local weather (e.g., stream water temperature) conditions. Science diplomacy across the four riparian states is needed to enable the development of Indus basin-wide eco-hydrological models that can improve the prediction of available water (and reduce uncertainty) under alternate climatic and land-use scenarios, which in turn can provide vitally important information to all the relevant agencies in the riparian countries for improving the local/community and regional scale planning for ensuring water, energy, and food security.

Both Neelum-Jhelum river and Kabul river tributaries of the Indus (Fig. 7.1) provide unique experimental opportunities for track-2/science diplomacy. While mostly neglected in the literature on transboundary environmental cooperation and diplomacy, both of these river systems illustrate the problem of forecasting precipitation and other eco-hydrological factors without field data. High-resolution data sharing across these two river systems can improve the capabilities of the current hydrological models in the three countries that would generate win-win-win scenarios for all three and could proactively avoid worst-case flooding events as recently experienced in Kashmir and lower Kabul delta.

7.5 ICIMOD's Role Thus Far

Interestingly, cooperation on environmental matters in the region predates the establishment of SAARC in 1985. The unique characteristics of the Himalayan region, featuring the world's highest mountain range with the steepest elevation gradient, prompted the creation of the International Centre for Integrated Mountain Development (ICIMOD), which has its roots in the United Nations Educational, Scientific, and Cultural Organization (UNESCO) "Man and the Biosphere" program, launched in 1983. The government of Nepal offered to host the new institution, and the governments of Switzerland and Germany, along with UNESCO, agreed to act as the founding sponsors. Nepal and UNESCO signed the formal agreement and inaugurated ICIMOD in December 1983 with its headquarters in Kathmandu, legitimized through an act of Parliament in Nepal the same year.

Fig. 7.1 Overview of river systems in the Indus basin across Pakistan, India, China, and Afghanistan

The center has the formal mandate "to enable and facilitate the equitable and sustainable wellbeing of the people of the Hindu Kush Himalayas by supporting sustainable mountain development through active regional cooperation."[9] ICIMOD is governed by a board of governors comprising one representative from each of the regional member countries and independent members who are nominated by the ICIMOD support group based on their recognized professional expertise and experience. One of the regional cooperation initiatives of ICIMOD has been the establishment of the Himalayan University Consortium (HUC) for Mountain Development Studies, which has the stated goals:

[9] International Centre for Integrated Mountain Development (ICIMOD). Mission statement online: http://www.icimod.org/?q=122

To promote and support the conducting, acquiring, preserving, and sharing of mountain research and to develop data, information, and knowledge through academic and nonacademic means and platforms
To provide open and equitable access to these knowledge resources to members and possibly others in the region
To promote the effective use of the available knowledge through training courses, academic curricula, student/faculty exchanges, and web-based information portals

As the largest country in South Asia, India is perceived by its neighbors to assume a particular responsibility for action on regional ecological concerns. In 2008, Indian prime minister Manmohan Singh announced that India would pursue eight national "missions" for sustainable development: developing solar energy, enhancing energy efficiency, creating a sustainable habitat, conserving water, preserving the Himalayan ecosystem, creating a green India, creating sustainable agriculture, and, finally, establishing what he called a "strategic knowledge platform for climate change." In announcing these missions, Singh noted that India traditionally has treated nature "as a source of nurture and not as a dark force to be conquered and harnessed to human endeavor. There is a high value placed in our culture to the concept of living in harmony with nature" (Timmons 2008). Yet the promise of this plan was tempered by domestic concerns regarding the pace of development, and a few months after the launch of the national missions, the external affairs minister noted that "political compulsions force us to meet the aspirations of our people quickly even as we subject ourselves to newer and more rigid international standards and norms" (Mukherjee, 2008).

Within this national platform, however, the area where regional cooperation should be further explored is in the mission for a "strategic knowledge platform for climate change." In addition, there is the Climate and Development Knowledge Network (CDKN), which was established under the Copenhagen Climate Change Summit in 2009. The regional South Asian hub of this network is based at a Pakistani NGO called Leadership in Environment and Development (LEAD-Pakistan).[10] India's Energy and Resources Institute (TERI) and LEAD-Pakistan cohosted the first joint deliberative program of the two countries' planning commissions in 2013.

Environmental cooperation between Afghanistan and Pakistan could also play a role in more regional stability. For example, data sharing and technical cooperation could eventually pave the way toward a bilateral Afghan-Pakistani water resources commission and perhaps even a treaty governing the Kabul River's resources. Pakistan sent a technical committee to Afghanistan in 2003, and the World Bank in 2006 offered support for joint consultations. Yet distrust between Afghanistan and

[10]The LEAD network of nongovernmental organizations was initiated by the Rockefeller Foundation in 1995 to create and sustain a global network of leaders who are committed to promote change toward patterns of sustainable development that are economically sound, environmentally responsible, and socially equitable, in line with Agenda 21 promulgated at the United Nations Conference on Environment and Development (UNCED: Rio Summit, 1992).

Pakistan has hampered such initiatives, though some tentative efforts continue (Renner 2012).

The 1997 Convention on the Law of the Non-Navigational Uses of International Watercourses finally entered force in 2014 but has been to date ratified by less than 20 % of the United Nations' member states. The lack of an international consensus on water continues to be a concern for using a treaty-based approach to regional "hydro-diplomacy." Five environmental treaties that were promulgated by the UN Economic Commission for Europe could provide important models for South Asia as well. These five treaties are:

Convention on Long-range Transboundary Air Pollution
Convention on Environmental Impact Assessment in a Transboundary Context
Convention on the Protection and Use of Transboundary Watercourses and International Lakes
Convention on the Transboundary Effects of Industrial Accidents
Convention on Access to Information, Public Participation in Decision-making, and Access to Justice in Environmental Matters

Interestingly enough, the definition of "Europe" in this context extends as far as Tajikistan, which borders the Pamir region and Wakhan. It would be worthwhile to explore the extension of these agreements or their applicability to South Asia through the United Nations system. Extending this framework to South Asia is particularly significant, given the rise in extreme weather events and disaster response cooperation in this region.

The closest South Asia that has come to such a system was when the World Meteorological Organization (WMO) and partner countries, alongside ICIMOD, developed a disaster mitigation project titled the Hindu Kush-Himalayan Hydrological Cycle Observation System (HKH-HYCOS). The stated aim of this project is to "enhance regional cooperation in hydro-meteorological data collection and sharing for flood forecasting to support disaster prevention and flood management at the regional level." The project has established a regional flood information system (RFIS) to facilitate transboundary exchange of real and near real-time data, best practices, and know-how in support of flood management. It also seeks to build the technical capacity of the national hydrological and meteorological services of partner countries. The overall objective is to mitigate casualties and property damage through timely exchange of flood data and information between and among partner countries.

The project was initiated in May 2001 with the financial support of the US State Department Regional Environmental Office for South Asia and the US Agency for International Development (USAID) Office of US Foreign Disaster Assistance (OFDA), based on the proven concept of WMO's World Hydrological Cycle Observing System (WHYCOS). Technical experts and government representatives from partner countries were invited to the first phase in a deliberative process. The current phase, which started in December 2009, is supported by the Finnish government's Ministry of Foreign Affairs (See Fig. 7.2).

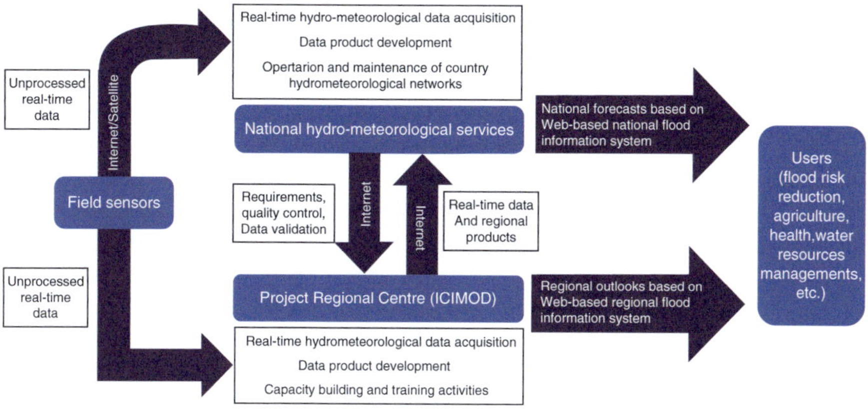

Fig. 7.2 Regional cooperation on flood control in South Asia (ICIMOD, WMO 2012)

7.6 Limitations of SAARC' s Water Mandate

Much as the Indus Waters Treaty (IWT) is celebrated in terms of preventative "hydro-diplomacy," the treaty employed a rather linear process of dividing up the major tributaries of the Indus between India and Pakistan and providing some risk assurance through the construction of dams. Ecological factors in terms of arability impact on land and biodiversity were not on the horizon of negotiators in 1960, when the treaty was finalized. It was thus not meant to be a truly "cooperative" agreement but rather a symptomatic response to prevent further escalation of conflict.

Clearly there are some distributional aspects of water-sharing between India and Pakistan that are inherently a "zero-sum game." In such cases there is little prospect for trying to use ecological factors as a means of improving trust. The dispute resolution procedures within the treaty should be continued and perhaps refined to cope with such challenges. The Baglihar dam dispute resolution of 2007, which involved Pakistan's challenge of an impoundment on the Chenab River, exemplifies the scope and limitations of the IWT mechanism (Salman 2011). The arbitrator resolved the dispute in 2007 with allowance for gated spillways for India and a reduction in the height of the proposed dam to satisfy Pakistan's concerns, purely on engineering feasibility criteria that met the water distribution mandate of the treaty, rather than broader watershed sustainability considerations. The Kishanganga arbitration was perhaps more innovatively settled in December 2013 with both India and Pakistan claiming some level of victory in the matter.[11] The ongoing Tulbul case also opens

[11] For an analysis of this arbitration, refer to Prakash Pilla's commentary *Kishenganga arbitration and viability of arbitration in resolving State-to-State disputes*. Published online at multiple venues – July 17, 2015: http://kluwerarbitrationblog.com/2015/07/16/kishenganga-arbitration-and-viability-of-international-arbitration-in-resolving-state-to-state-disputes/

up the issue of tourism development, which could be a cross-border cooperative activity if appropriately configured. In this case, India is planning the controlled release of water during the drier months of October to February to facilitate navigation for trade and tourism; but this case will most likely require arbitration because a broader regional benefits framework has not been presented.[12] India maintains that the regulating structure is permitted under the IWT for the nonconsumptive use of navigation and that no "pondage" of water was being created over the Jhelum River. Pakistan, however, contends that this is a "storage project" and will deprive Pakistan of its rights under the treaty. With growing scarcity of water due to climate changes as well as demographic pressures, such disputes are likely to escalate. The treaty does not have a cooperative mechanism predicated on ecological premises and in some instances has resolution mechanisms that are inherently nonadaptive to environmental change.

For example, the allocation of the Chenab and Jhelum rivers to Pakistan still provides for a fixed "top-up" of water for India from these rivers every year that is approximately enough to irrigate 320,000 ha of cropland . Another recent dispute that highlights the limitations of the treaty involves the use of hydroelectric dam construction to claim carbon credits. India managed to secure carbon credits for the Nimoo-Bazgo project, which has been contested by Pakistan since 2002; Pakistan also claims India did not respond to its request for engagement on the matter until December 2006. The Nimoo-Bazgo hydroelectric project, with a capacity of 45 MW, has been built near Alchi village in India's Leh district. As this dispute shows, although climate change could be a catalyst for cooperation in terms of collective planning for extreme weather events, it can also create incentives for new infrastructure projects that push the limits of the current riparian sharing arrangements.[13]

While there is little doubt that the Indus Waters Treaty was not structurally designed to deal with climate change uncertainties, views differ on whether renegotiating the treaty is the proper path to preventing conflict over water. Kugelman (2012) has argued that "the best hope for averting water war lies not in repairing frayed political ties or enhancing water diplomacy, but rather in better managing domestic water resources." He contends that infrastructure that was the outgrowth of the treaty was not efficiently designed and that domestic policies have not been adequately maintained or adapted to cope with climatic and demographic stresses. Thus water conservation technologies and the repair of leaky pipes and canals, as well as cooperative frameworks, could be a more assured path to conflict mitigation than a renegotiation of the treaty itself. Sharing science and engineering expertise to enhance such conservation strategies could still be a cooperative mechanism between India and Pakistan and indeed other regional partners. Such cooperation

[12] The International Court of Arbitration process that decided on the Kishanganga dispute issued an order in October 2012 allowing a Pakistani delegation to inspect the site of the dam and visit the Wullar Barrage and Tulbul navigation projects (Parsai, 2012).

[13] "India's move to fill Nimoo-Bazgo dam in J&K irks Pak." Rediff News, August 12, 2012: http://www.rediff.com/news/report/india-s-move-to-fill-nimoo-bazgo-dam-in-kashmir-irks-pak/20120812.htm

could be more easily framed within the mandate of SAARC rather than trying to renegotiate the treaty, which is excluded from the SAARC charter's mandate.

Just as water is being lost through bad infrastructure, energy losses due to inefficient power lines is an immense challenge in the region. Much of the impetus for developing water infrastructure in the region comes from the growing need for power in South Asia. The rapid rise of population and the pressures for industrialization are leading to an unprecedented demand for energy. Thus water infrastructure policy will need to be tied to decision-making on multiple sources of energy. Such connections are, however, very tenuously made. For example, while developing large dam infrastructure primarily for power (as stated by India in the construction of numerous "run-of-the-river" projects) has led to further conflict, the development of natural gas pipelines has the potential for a more cooperative infrastructure outcome if managed with appropriate measures (Ali 2010). The Energy Charter, which is a treaty-based agreement for improving international cooperation on energy trade and transit and providing for a dispute resolution mechanism (established in 1991), has thus far been limited in its efficacy. It could be used to strengthen a proposed protocol on transboundary pipelines.[14] Natural gas is an important transition fuel and more versatile than hydropower, as it can be used for mobile transport and as a direct fuel for heating and cooking (thus making it more efficient in most uses).

In addition, renewable energy sources in terms of small-scale hydropower, solar, and wind should still be given importance for rural electrification despite their limitations in meeting large-scale demand. Networks such as the USAID-funded South Asia Regional Initiative for Energy (SARI/Energy) are important efforts in this regard. The mandate of SARI/Energy is to "promote technical and institutional frameworks for regional energy planning and infrastructure investment involving cross-border trade in energy."[15] This is a bold effort with the potential for considerable impact in fostering broader linkages between water and power. However, here too, USAID has been limited in its ability to more directly encourage Indo-Pakistan cooperation, given the lack of political will on the part of the US administration to make such leveraging a priority. Such cooperative efforts are likely to succeed only if there is recognition among the countries of the region that river systems and energy sources have fundamental ecological underpinnings and should not be narrowly defined by political or ethno-religious boundaries.

An interesting development in riparian cooperation was the establishment in 2000 of the Mekong-Ganga Cooperation (MGC) initiative during a ministerial

[14] As noted on the Energy Charter Web site accessed August 14, 2015. "Taking into account the position of the EU, the Charter Conference decided on 29 November 2011 to repeal the negotiation mandate of 2009. In view of the possibility of a reset of negotiations on a new Protocol, the Trade and Transit Group was tasked to conduct consultations among ECT members (in 2012), observers and industry representatives, in order to obtain more information with regard to the prospects for such an initiative, including the issues to be addressed and the convergence of positions within the constituency in this regard."

[15] Mission statement from SARI/Energy Web site, accessed August 14, 2015: http://www.sari-energy.org/

meeting of six member countries: India, Thailand, Myanmar, Cambodia, Laos, and Vietnam. The countries emphasized four areas of cooperation between the river basins—tourism, culture, education, and transportation linkage—as a solid foundation for future trade and investment cooperation in the region. The MGC initiative uses a riparian frame to focus on cultural cooperation among countries believed to have been influenced by Indic culture. The organization's exclusion of Bangladesh, which has much of the Ganges delta within its borders, shows that the initiative did not have ecological origins, despite using rivers as the locus of inclusion. However, following the sixth ministerial meeting of the MGC in New Delhi in September 2012, the Bangladeshi prime minister raised the issue of the country's exclusion during conversations with the Vietnamese leadership.[16] River systems and their potential for energy and cultural connectivity are also inextricably tied to the region's agrarian economic roots. Cooperation on land use for food crops and their linkage to human security deserves far greater attention as well.

High-resolution hydrometeorological models that account for nonstationarity[17] induced by global climate change at the global scale and land-use/land-cover change at the local scale are direly needed for all the river systems comprising the Indus basin. More intense monsoon precipitation events coupled with deforestation land-use scenarios would have different likelihoods of flood compared with a forest conservation scenario in the upper catchments of Indus basin. Accounting for uncertainties in glacial melt timing due to global warming or changes in the direction and timing of monsoon storm systems requires development of a lot better (both high-resolution and dynamic) eco-hydrological and/or hydrometeorological models. In a recent study of the upper Indus (Hunza and Gilgit river systems), Akhtar et al. (2008) used the PRECIS model at 25×25 km resolution and drove it with statistical downscaling of GCMs to run different glacial melt scenarios under different climate change scenarios. They concluded that statistical downscaling needs to be replaced with a dynamic regional model (such as NCAR's WERF model) for driving more high-resolution (sub-KM level) models that could potentially generate more accurate flooding and drought forecasting information. Dynamic downscaling of WERF model would require calibration of the model with high-resolution sensor and gauge databases across the administrative boundaries in conflict zones (e.g., Jhelum-Neelum and Kabul). Essentially, this technical process entails more computational complexity and hence greater data resolution (which will also add to the expense of the system).

More coarse scale studies at the Indus basin level (compared with other big river systems such as the Yellow and Brahmaputra) have generated dire warnings about the water and food security scenarios expected in the Indus in the face of climate change. Immerzeel et al. (2010:1384), for example, found

[16] "Hanoi's support sought for Dhaka's entry." The News Today, Dhaka, November 2, 2012: http://www.newstoday.com.bd/index.php?option=details&news_id=2327818&date=2012-11-03

[17] For descriptive details on this technical concept, refer to World Meteorological Organization's factsheet online http://www.wmo.int/pages/prog/hwrp/chy/chy14/documents/ms/Stationarity_and_Nonstationarity.pdf (Accessed, November 6, 2015).

> Upstream water supply is crucial to sustain upstream reservoir systems, which are used to store and release water to downstream areas when most needed. Irrigation water for the Indus Basin Irrigation System, which is the largest irrigation network in the world, is, for example, regulated through two major storage dams (Tarbela dam on the Indus River and the Mangla dam on the Jhelum River). Both are located in the upper Indus basin and are fed predominantly by meltwater. Any change in upstream water supply to these dams will have a profound effect on millions of people downstream. Our results show a substantial variation in changes in future water supply.

Further, Immerzeel et al. (2010: 1385) concluded that

> Regardless of the compensating effects of increased rainfall in the two basins with the largest NMI [Normalized Melt Index], the Indus and the Brahmaputra, summer and late spring discharges are eventually expected to be reduced consistently and considerably around 2046–2065 after a period with increased flows due to accelerated glacial melt….. The effects in the Indus and Brahmaputra basins are likely to be severe owing to the large population and the high dependence on irrigated agriculture and meltwater.

Improved high-resolution characterization of these scenarios that could be used for early warning systems (e.g., Zia and Wagner-Hammond 2015), flood zone planning (e.g., Zia and Glantz 2012), and drought/famine management is needed through more dynamic and integrated models. Improved estimation of the risk posed by complex climate change, land-use change, and political change scenarios can be very useful in ensuring food, water, and energy security in the Indus basin. Recent developments in transboundary collaboration across Amazon countries (such as Brazil, Peru, Ecuador, and Columbia) and in the Middle East (in particular Jordan and Israel) provide a successful example of science and environmental cooperation that can be emulated in observing and managing the risk in the Indus basin.

7.7 Conclusion

Since the ascent to power of Narendra Modi in India in May 2014, prospects for an Indo-Pak détente have faltered. The economic pragmatism of Modi's past policies was believed by some analysts to trump ideological concerns. Environmental factors have not figured prominently in Mr. Modi's agenda. However, at the SAARC summit in November 2014, the signing of the SAARC Framework Agreement for Energy Cooperation suggested that prospects for hydro-diplomacy might be improving. However, because of mutual distrust and local political concerns, neither India nor Pakistan has been able to operationalize this framework thus far. As suggested in this chapter, it is essential to link such conversations to regional institutions and to focus on the salience and mutual benefits of ecological data sharing, particularly with reference to natural disasters. The Nepal and Badakhshan earthquakes of 2015 have once again highlighted the extreme vulnerability of this region to natural cataclysms. Security concerns around such sharing could be alleviated through multilateral frameworks such as the Sendai Framework for Disaster Risk Reduction.[18] However, these processes themselves need to consider regional nuances and computational tools to make their operational plans more meaningful.

The analysis presented in this chapter has attempted to show how data sharing has some limited historic precedence in South Asia and where strides are needed to ensure environmental security of the region. The role of existing regional cooperation organizations should not be underestimated in this regard. Although organizations such as SAARC have been largely ineffective in matters of high politics such as broader territorial conflict resolution, they have much potential for ecological peace-building. Knowledge-building organizations such as ICIMOD have also played an important role in laying the groundwork for such broader attempts at environmental diplomacy. Transboundary data sharing should be mainstreamed within South Asia if the regional planning vision of "Industan" is to reach fruition.

References

Ahmed ZS (2013) Regionalism and regional security in South Asia: the role of SAARC. Ashgate Publications, Oxford

Akhtar M, Ahmad N, Booij MJ (2008) The impact of climate change on the water resources of Hindukush–Karakorum–Himalaya region under different glacier coverage scenarios. J Hydrol 355(1):148–163

Ali SH (2010) Energizing peace: the role of oil and gas pipelines in regional cooperation. Brookings Doha Centre, Qatar

Ali SH (2012) Siachen tragedy: an opportunity for peace. National Geographic Newswatch, April 7, 2012 http://voices.nationalgeographic.com/2012/04/07/siachen-glacier/

Ali S et al. (2015) Twenty first century climatic and hydrological changes over upper Indus Basin of Himalayan Region of Pakistan. Environ Res Lett, v. 10

Bagla P (2006a) Across a political divide, researchers converge on Himalayan plan. Science 313(5783):30–31

Bagla P (2006b) Pakistan gives geology conference the cold shoulder. Science 312(5777):1117–1117

Desai VV (2010) Political economy of regional cooperation in South Asia, Manila, Philippines, Asian Development Bank working paper series on regional integration, No. 43, July, 2010

Hijioka et al. (2014) Change, intergovernmental panel on climate. 2014. Climate Change 2014 – impacts, adaptation and vulnerability: part B: regional aspects: volume 2, regional aspects: working group II contribution to the IPCC fifth assessment report. Cambridge University Press. 2014

ICIMOD, WMO (2012) Establishment of a regional flood information system. ICIMOD and WMO, Kathmandu, http://www.icimod.org/resource/19357

Immerzeel WW, Van Beek LP, Bierkens MF (2010) Climate change will affect the Asian water towers. Science 328(5984):1382–1385

Kugelman M (2012) Repairs could stifle South Asia's water war. Global Times, October 11, 2012

Mukherjee P (2008) On India and global challenges: climate change and energy security. Presentation at the Asia Society, New York, September 30, 2008

[18] For further details, refer to the Sendai Process web page: http://www.wcdrr.org/preparatory/post2015

Parsai G (2012) India for arbitration of Tulbul row. The Hindu, March 30, 2012: http://www.thehindu.com/news/national/article3259314.ece
Pomfret R (1997) The economic cooperation organization: current status and future prospects. Eur Asia Stud 49(4):657–667
Pyatt G, Delivering the New Silk Road. Tokyo, Japan, July, 9, (2012). Accessed online 2 Aug 2015, from U.S. Department of State official site: http://www.state.gov/p/sca/rls/rmks/2012/194735.htm
Renner M (2012) Water as a transborder problem of Afghanistan. In: Riecke H (ed) Partners for stability: involving neighbors in Afghanistan's reconstruction – transatlantic approaches, DGAP Schriften zur Internationalen Politik. Nomos Verlag, Baden-Baden
Reuters. "India, Pakistan to join China, Russia in security group," Reuters, July 10, 2015: http://www.reuters.com/article/us-china-russia-idUSKCN0PK20720150711
SAARC (1992) Regional study on the causes and consequences of natural disasters and the protection and preservation of the environment. SAARC Publications, Kathmandu
Salman SM (2011) The Baardhere Dam and water infrastructure project in Somalia—Ethiopia's objection and the World Bank response. Hydrol Sci J 56(4):630–640
Salman M, Salman A (2008) The Baglihar difference and the resolution process: a triumph for the Indus waters treaty? Water Policy 10:105
SANDEE (2011) A decade in focus. South Asian Network for Development and Environment Economics, Kathmandu
Timmons H (2008) India announces a climate change plan. New York Times, June 30, 2008
Wang F (2011), Grand strategy in the great game–strategic interests and objectives of states of the Shanghai Cooperation Organization. Burlington, Vt.: Institute for Environmental Diplomacy & Security at the University of Vermont. http://www.uvm.edu/ieds/sites/default/files/SCO_casestudy_V4.pdf. Accessed online 1 Sept 2015
World Meteorological Organization (2012) Establishment of a regional flood monitoring system. ICIMOD Publications, Kathmandu
Zia A (2013) Post-Kyoto climate governance: confronting the politics of scale, ideology and knowledge. Rutledge, London
Zia A, Glantz M (2012) Risk zones: comparative lesson drawing and policy learning from flood insurance programs. J Comp Policy Anal Res Pract 14(2):143–159
Zia A, Wagner CH (2015) Mainstreaming early warning systems in development and planning processes: multi-level implementation of Sendai Framework in Indus and Sahel. Int J Dis Risk Sci 6(2): 189–199. doi:10.1007/s13753-015-0048-3

Chapter 8
The Indus Basin: The Potential for Basin-Wide Management Between India and Pakistan

Douglas Hill

Abstract This chapter is concerned with how water can be sustainably managed across the Indus basin, focusing specifically on the challenges of surface water management in India and Pakistan. Governments in both countries have laid emphasis on the expansion of hydropower for storage, irrigation, and energy trading, and consequently the politics of water is increasingly implicated in the geopolitics of the Indus basin. Within each country, the control of water varies across time and space and reflects the dynamics of broader power structures such as those related to the relationship of the state to different social groups. The supply-side hydraulic paradigm that has historically predominated in both countries engenders a technocratic institutional culture and securitized discursive environment that is resistant to the voices of nonelite actors. As the private sector has assumed a greater role in hydropower construction, particularly in the Indian Himalayas, there has not been a concomitant shift in the transparency or accountability of water governance institutions, as advocates of economic liberalization suggest there should be.

Clearly, then, accommodating the needs of different stakeholders beyond the status quo requires a challenge to prevailing state-society relations within both India and Pakistan at a variety of scales. In turn, effective basin-wide management will require a transformed institutional culture that is more open to polycentric formations. In arguing that such a change in institutional culture is both desirable and possible, the latter part of the paper highlights the role that multitrack dialogues, education, and media can all play as part of the promotion of a de-securitized basin where peaceful and sustainable relations replace long-standing conflict.

Keywords Indus basin • Multitrack dialogues • Transboundary • Chenab • Hydropower • India • Pakistan • Jammu and Kashmir • Himachal Pradesh

---

D. Hill (✉)
Department of Geography, University of Otago, Dunedin, New Zealand
e-mail: dph@geography.otago.ac.nz

Z. Adeel, R.G. Wirsing (eds.), *Imagining Industan*, Water Security in a New World, DOI 10.1007/978-3-319-32845-4_8

8.1 Introduction

This chapter is concerned with how water can be sustainably managed across the Indus, focusing specifically on the challenges of surface water management in the Indian and Pakistani portions of the basin. Uneven development, pollution, growing water and energy insecurity, institutional rigidity, and demographic pressures all create difficulties in managing these resources. In outlining these pressures and assessing why they have become so challenging, this chapter begins from the premise that the control of water in the Indus is embedded in the contestation between different parts of society and thus must be understood as the result of the interplay between biophysical resources and broader political, social, and cultural factors (Mollinga 2008). In both countries, institutions of the state remain biased toward particular classes and regions, with the effect that water resource development in both India and Pakistan reinforces elite interests in dominant provinces at the expense of other groups. Thus how water is distributed, and the dominant meanings attached to its usage, can be understood with reference to dynamics of larger power structures predominating in different parts of the Indus basin, such as those related to the relationship of the state to different social groups. These state-society relationships can be understood on the basis of class, caste, ethnicity, and gender dynamics and vary across time and space. Development, however one might choose to define it, is spatially and socially highly uneven across the Indus basin, with seemingly intractable poverty traps and high levels of prosperity both evident.

Furthermore, previous chapters have left no doubt that the resultant dynamics of water contestation in the Indus create impediments to institutional change, the trajectory of which is extremely difficult to alter in the short to medium term. As well as these issues of state-society relations, water governance in India and Pakistan continues to be beset by significant bureaucratic fragmentation, with different subjects handled by different departments, as well as by numerous bodies of legislation in both countries, related to minor and major irrigation, drainage, forestry, and environmental protection. Further, the fact that the Indus is effectively a closed basin where, in biophysical terms, there is limited capacity to increase allocation means that policymakers are working in an extremely constrained policy space. Decisions over allocation are, in turn, increasingly securitized, which has the effect of delimiting debates so that they are overwhelmingly concerned with a narrowly defined "national interest" framed in terms of the resolution of technical issues. The heightened tensions over transboundary water sharing between India and Pakistan are increasingly folded into broader issues of Kashmir, terrorism, and cross border skirmishes. This creates a restricted discursive environment that is resistant to the voices from other stakeholders (Hill 2015; D'Souza 2014).

Previous chapters in this book have foregrounded this perspective and have outlined the range of social, cultural, political, and economic challenges that problematize effective basin-wide management. The problematic role of the state in managing water resources has emerged as a significant issue in both countries. Furthermore, the terms of debate over how the waters of the Indus should be utilized are further restricted by the fact that development is often seen as synonymous with economic growth, and so priority is afforded to those schemes that are seen to enhance this, even if in the process they simultaneously generate a range of other problems. Indeed, a key argument of this chapter is that the growing pressure for the wide-scale transformation of the Hindu Kush Himalayas through the building of a great many large- and medium-scale dams (Hill 2013a, b; Pomeranz 2009) is not within itself a way to resolve the difficult challenges associated with the food, water, energy, and environment nexus in the Indus basin.

In assessing the challenges and opportunities engendered by the expansion of hydropower in the Indus, this chapter argues that, as it currently stands, the acceleration of large-scale dam construction in the Indus basin will intensify the negative consequences that we have seen in the past, both in terms of sustainable development and human security. This is because the regulatory framework for hydropower expansion in both Pakistan and India is far from sufficient for the task, particularly with regard to the cumulative impacts of economic integration through hydropower and energy trading. This will impact upon the livelihoods of communities and on the region's biodiversity in ways that are insufficiently understood or considered by those who are enthusiastically promoting this model of development. Thus, despite the potential for closer economic ties to engender closer relationships and perhaps promote a more politically open basin, it must also be acknowledged that this kind of infrastructure-led development in South Asia has often inflamed tensions between Pakistan and India as well as between federal and state governments (Hill 2009). The chapter illustrates these broader arguments with reference to examples drawn from the Indian states of Himachal Pradesh and Jammu and Kashmir, both of which have contentious hydro-politics.

While mindful of the difficulties in changing the current practices of water resource governance in the Indus, this chapter also joins others found throughout this volume in seeking to chart a way through these obstacles, suggesting these resources can be managed sustainably through increasing transparency, ensuring more robust participation, and broadening the range of those who benefit from development. The contentions over water are amplified and become embroiled in broader political issues. As such, rebalancing the needs of a range of different stakeholders at a variety of scales requires a challenge to prevailing state-society relations. In turn, effective basin-wide management will require a transformed institutional culture that is more open to polycentric formations and can move beyond the singular focus upon a supply-side hydraulic paradigm that has historically predominated in the rigid bureaucracies of India and Pakistan.

In arguing that such a change in institutional culture is both desirable and possible, the latter part of the paper highlights the role that multitrack dialogues, education, and media can all have as part of a broader promotion of a more peaceful and sustainable set of relations in regions beset by long-standing conflict. Thus, while there is clearly a range of constraints associated with transboundary cooperation over water resources at a variety of scales, it is also the case that there are processes and mechanisms by which sustainable development can be engendered in this contested region.

8.2 Challenges of Water in the Indus Basin

As previous chapters have outlined, people in the Indus basin face a significant number of challenges related to declining water availability per capita, both in terms of water quality and quantity. Agriculture remains the most important usage of water and the major contributor to the economic base of the basin, both as a source of subsistence food security and rural livelihoods and as an income-generating activity (Ahmad 2012). Thus, those living in the mountainous parts of the Indus basin continue to rely upon monsoon-variable, rain-fed agriculture and tend to be economically and politically marginalized, whereas in the more prosperous parts of the basin, notably in both parts of Punjab, agriculture generates high surplus returns and has provided the opportunity for diversification into nonagricultural industries. Many of the most important industries in the Indus basin are very water intensive, including textiles, sugar, and wheat, and their continuing viability on the Pakistani side in particular is problematized by the continuing low efficiency of irrigation, technological obsolescence, poor regulatory mechanisms, and worsening power shortages (Briscoe and Qamar 2006). A changing climate intensifies the vulnerability of people in the Indus basin to these challenges (Lieven 2012), with "too little and too much water" (Pradhan et al. 2012) creating a cycle of droughts and floods that has already had major impacts on livelihoods and looks to increase in the future. Moreover, while surface and groundwater irrigation have increased enormously in the period since the Green Revolution began in the early 1960s, a considerable decline in both quality and quantity of the groundwater resource is now evident (Rodell et al. 2009). While there are variations across the Indus basin, this situation holds true for both India and Pakistan, making the agricultural base, from which the Punjab's prosperity has been derived, evermore tenuous.

While all of these issues ostensibly appear to be mainly questions of limited water supply being outstripped by the multiple demands of a growing population, in reality those who control water in the Indus and the purpose for which they use it are inherently driven by social, economic, and political factors (Mustafa 2010). At the heart of the challenge of managing water across the Indus basin is the overcoming of poor governance on both sides of the border (Hill 2013a, b; Mustafa et al. 2013; Lahiri-Dutt and Wasson 2008; Briscoe and Qamar 2006; Lahiri-Dutt 2000; Roy 1999). Numerous authors have extensively outlined the systemic failures of the high modernist ambitions of the state, with criticism frequently aimed at the lack of responsiveness from the government with regard to appropriate procedures for the mitigation of social, economic, and environmental impacts. For example, in assessing governance challenges in Pakistan, Mustafa et al. (2013, 21) argue that the state there displays "a seemingly general apathy towards changing current patterns lack of regulatory oversight and seemingly endemic corruption."

A range of authors have analyzed and documented how the hydraulic state has consistently represented the interests of dominant classes in both India and Pakistan (Hill 2009; Roy 1999). In the work of Imran Ali (2004), for example, there is a clear demonstration of how the historical trajectory of the canal colonies from the colonial to the postcolonial period in Pakistan has intensified the control of Punjabi landlords, the military, and the bureaucracy over other social groups. To Ali, the control and distribution of water demonstrate how the state is an instrument of control for these classes, and as such it is unsurprising that governance issues continue to bedevil the sector in ways that solidify their dominance within society.

Thus while it is certainly the case that Pakistan has a relatively low volume of water available per capita, an environmentally deterministic or neo-Malthusian explanation does not do justice to the institutional context that supports the promotion of the interests of a class of Punjabi large landholders, who themselves have strong connections with the bureaucratic and military apparatus of the Pakistani state (Ali 2004). In a similar way, the prosperity of Punjabi and Jat farmers is underwritten by the minimum support prices of the government of India for the purchase of wheat and the highly subsidized electricity (and diesel) that enables the extraction of groundwater (Hill 2003). In contrast to this state-enabled prosperity, it is not simply the topography but also the economic structure and political marginalization of parts of the Indus, such as Azad Kashmir or Ladakh, which contribute to the fact that the gains from the development of the Indus basin are extremely uneven in terms of both spatial distribution and social relations.

An important factor in controversies over water sharing in the Indus basin is disagreement over the appropriate scale and distribution of hydropower development. While large-scale dams can transfer surface water to areas of greatest demand at times when it is most needed, offer flood protection, and be an important source of electricity, too often the gains from their construction in South Asia have not been enough to offset the negative impacts that they have engendered (D'Souza 2008; Hill 2008). Displacement and disruption of livelihoods for people living in the immediate environs of these dams have often been accompanied by impacts on both riverine and riparian ecosystems that have impacted people living downstream (McCartney et al. 2001). As well as being controversial in terms of loss of livelihoods and ecosystem services, hydropower has been a long-standing source of political contestation at regional, national, and local scales. This is unsurprising considering that rivers originating in Tibet and the Hindu Kush Himalayas drain into some of the most politically fragmented parts of Asia, including the upper reaches of the Indus. Further, the allocation of water continues to be implicated in interprovincial politics, as we find with the Sutlej-Yamuna Link Canal dispute in India or the Tarbela Dam in Pakistan (Hill 2009).

At the all-basin scale, the allocation of water is conditioned by the fact that the 1947 partition of the Indian subcontinent divided the eastern and western rivers between India and Pakistan (Gazdar 2005). The regulatory framework for this division was set out through the Indus Waters Treaty (IWT) and the appointment of the Permanent Indus Commissioners to resolve any disputes or differences that arose between the two countries. While widely regarded as successful (Salman and Uprety 2002), in recent years many commentators have been asking for a renegotiation of the IWT (see Magsig, this volume). Contention over the legality and appropriateness of a range of individual projects has intensified in recent years (Hill 2013a).

Disputes over water resources have different dimensions on either side of the basin. In India, the harnessing of these rivers is projected to ease development of marginalized provinces, particularly the troubled Jammu and Kashmir province; however, a long-standing collection of civil society groups reject these claims as not being borne out by the history of dam building both in Kashmir and elsewhere in the country. While the hydropower industry in India has not approached anything near "best practice" when it comes to anticipating and mitigating negative social, political, and environmental impacts, it is certainly true that civil society has been relatively successful in slowing down the expansion of the industry. However, it is clear that the new National Democratic Alliance (NDA) government under Narendra Modi has a strong interest in expanding the harnessing of the Himalayan rivers. Indeed, that Modi's first foreign policy engagements were to Bhutan and Nepal demonstrated that the government intends for hydropower to be a significant component of its strategy for energy security and regional integration (Hill 2015). The subsequent strong actions against NGOs that were seen to be impeding development, including those campaigning against energy projects, also suggest that this government believes that India's rise as a regional superpower may need to be facilitated through the silencing of dissent (Hill 2015; Mazoomdar 2014).

In Pakistan, many groups use the issue of water sharing with India to create a common enemy, including militant groups such as Jamaat-ud-Dawa, who have sought to increase their constituency by continual assertions about India's intention to use water as a strategic weapon. This continues to be a point of mobilization, even if the Indus River System Authority (IRSA) itself admits that India is not causing water shortages in Pakistan (News International 2015). At the same time, the lack of transparency in Pakistan's water sector is even more severe than it is in India, so that stakeholders from civil society are rarely given the opportunity to participate in a constructive or meaningful fashion about how the sector should be developed. Even so, large-scale hydropower development has been so mired in controversy in Pakistan that the expansion of the sector has been stalled.

8.3 Contesting Hydropower in the Indus Basin

When thinking through the manner in which the transboundary water resources become securitized in the Indus basin, it is clear that the terms of the debate about the costs and benefits of hydropower are frequently reduced to a binary representation of "Indian" interests versus "Pakistani" interests. The national interest of the nation-state must therefore be defended against the aggressive "other" who is intent on stealing the "nation's water." In this rendering, the national resource of water is easily conflated with other issues of national interest. The portrayal of the Indus as indicative of a neo-Malthusian crisis of environmental security (*a la* Homer-Dixon 1999, 1994) is reflected in the number of pieces in the past few years which suggests that a water war in the region is imminent (Chellaney 2011). The securitization of the Indus basin has arguably become intensified in the last few years as China has become a larger player in the geopolitics of water in the Indus and elsewhere in South Asia (Kondapalli, this volume; Hill 2013a).

There are a significant number of issues with this kind of representation. Firstly, such a framing clearly elides a range of significant issues occurring inside the nation-state, including the differential impact of any planned developments between upper and lower riparian provinces within the same country (e.g., such as between Punjab and Sindh). Secondly, by confining the debate only to what happens to the watercourse, the relationship to broader ecosystem services is often simplified. This is particularly significant both for biodiversity and for people whose cultural or social relationship with the water from the basin may be broader than just withdrawals for irrigation. Thirdly, by limiting notions of what constitutes security to the capacity of the sovereign state to enforce its territorial integrity against other sovereign states, security is defined in a very reductive manner that elides the issues of human or nontraditional security, which are nevertheless extremely pressing for many people in the Indus basin.

8.4 Economic Growth and Modernization in the Indus Basin

The securitization of water sharing is further complicated by the fact that there are a large number of projects proposed to be built in both the Pakistani and, particularly, the Indian-held parts of the Indus basin. Advocates of large-scale dams argue that this transformation will increase economic growth that will in turn lead to development and eventually to rising prosperity. Some commentators assert that increasing the economic integration of India and Pakistan, including through the development of shared hydropower and energy projects and facilities, could become a catalyst for peace and sustainable development in the Indus basin. The market-centered discourse, upon which such development strategy is based, suggests that regional integration can lower the cost of trade and increase market access, which is purported to be particularly beneficial for economically marginalized people living in remote and landlocked areas (ADB and ADBI 2013; Glassman 2010; Wilson and Otsuki 2007). The same arguments are often made for regions within the Indus basin, although mostly this is within either India or Pakistan, with only a few studies making similar assertions for the integration across these countries.

Certainly, the expansion of such infrastructure can create new livelihood opportunities, augment storage and flood protection capacity, and enhance cooperation across the Indus through energy trading (Molden et al. 2014). As such, there are clearly significant gains to be made from the closer economic integration of the different parts of the Indus (Indus Basin Working Group 2013). Indeed, there is no question that the infrastructure-led, market-centered approach is one foreseeable way of integrating and governing the different parts of the Indus basin, and there are indications that relations are heading in that direction. Pakistan has moved toward granting India nondiscriminatory access, which goes some way toward reciprocating the MFN status India granted Pakistan a decade ago, and India has made suggestions that it would be willing to export 500 MW of power to Pakistan (World Bank 2014). Donors such as the World Bank have already granted loans to facilitate the increased interconnectivity of other parts of South and Central Asia, most notably through electricity transmission and trade, and have publically called for India and Pakistan to be part of an integrated regional grid for energy. Certainly, then, the preconditions for the creation of such a regional energy grid are beginning to take shape in the Indus basin.

However, it must also be admitted that there are significant obstacles to increasing economic connectivity in this way. The military establishment in Pakistan profits from its control over the allocation of electricity; and the government of India is reluctant to have its northern grid vulnerable to disruption in a situation of renewed conflict. Even if such a unified, pan-Indus grid was to eventuate, there are few guarantees that the infrastructure built to facilitate this economic integration would be done in ways that were just and sustainable to local populations. Indeed, history would suggest that the expansion of such infrastructure would more than likely replicate many of the poor aspects of hydropower governance that have occurred in the building of large-scale dams in the past. On the other hand, a growing awareness of the many trade-offs involved in hydropower among donors and some bureaucracies in South Asia suggests that there are some grounds for optimism. Given this mixed situation, is it reasonable to expect that new investment into the mountainous regions of the Indus for hydropower construction will be done in a manner that improves on this generally poor history? The next section argues that, on the basis of evidence of ongoing construction of hydropower projects, there are significant doubts.

8.5 Controversies in Indian-Controlled Parts of the Indus Basin

One subregion of the Indus basin where the expansion of hydropower has been particularly controversial in recent years is in the Chenab subbasin. In the Indian-controlled states of Jammu and Kashmir and Himachal Pradesh, there are over 60 projects planned for construction. It has been widely noted that the construction of projects in this subbasin has been controversial because of the perceived impacts that this may have for Pakistan (Briscoe 2010; Committee on Foreign Relations 2011). Indeed, Pakistan objects to many projects on the Chenab, including the 1,000 MW Pakal Dul, the 120 MW Miyar, and the 48 MW Lower Kalnai hydropower projects (Alam 2015).

The latest project to symbolize the discord is the 850 MW power project Ratle Hydroelectric Power on the Chenab River. This run-of-the-river project came under construction in 2013 under the previous United Progressive Alliance (UPA II) government and is designed to be completed on a build-own-operate-transfer (BOOT) basis. However, in August 2015 the Pakistani government made objections to the Indian-based Indus Waters Commissioner, demanding the appointment of a neutral expert to resolve whether India's plan for setting up lower spillways for sediment management should be adjusted to reduce the risk of flooding. India has asked for talks, but if they cannot be resolved through the two Permanent Indus Commissioners established under the IWT, then it can go to a court of arbitration, as has occurred with Baglihar and Kishenganga in the past (Economic Times 2015). The 850 MW Ratle project is just the latest in a long line of projects on the Chenab that have been politicized in Pakistan, accusing India of purposely creating a strategic weapon against its western neighbor.

However, the planned expansion of large-scale dams in Jammu and Kashmir has critics within India as well, reflecting the intranational dimensions of the politics of water in that country (Hill 2009). Seismologists suggest that building so many dams in a quake-prone zone is hazardous, particularly because a great deal of the Chenab River runs along a fault line. The building of these projects has also been embroiled in center-state politics, particularly with regard to the government of India's treatment of the troubled state of Jammu and Kashmir (J&K). In that state, there are long-running complaints about the role of the government of India-owned National Hydroelectric Power Corporation Ltd (NHPC), which is one of the most significant companies in the entire country in terms of hydropower construction.

One of the objections to the role of the NHPC relates to the proportion of revenue from its hydropower projects given to the state government of J&K, with critics arguing that this should rise to 25 % from the current 12 % (Shah 2015). Furthermore, to many people in J&K, the NHPC is directly responsible for the fact that the state continues to suffer significant load shedding, particularly during the winter, while also having high prices for electricity. Indeed the state government wants the NHPC to return to state control a number of hydropower projects, including Salal, Uri-I, and Dulhasti; and animosities are such that the state government is asserting that it will not give any more projects to the corporation, and this is creating difficulties in the Peoples Democratic Party-Bharatiya Janata Party (PDP-BJP) alliance (Early Times 2016).

For its part, the Union Power Ministry rejects the arguments of the J&K government. It asserts that hydropower projects are the responsibility of many different ministries, and as such the interministerial nature of the projects means there are too many financial and legal hurdles to return the projects. Further, it argues that the state government gains concessions, including a proportion of the total output as "free" power, every time a new project begins operating. As such, the most advantageous position for the J&K government to take is to encourage the NHPC to build more dams. Indeed, the Union Power Minister recently asserted that since the Indian Himalayas contains a range of sites in other states, too much agitation from J&K would result in NHPC looking elsewhere.

It is not only in the J&K parts of the Chenab that hydropower construction is controversial. In the adjacent state of Himachal Pradesh, civil society groups have been protesting that a large number of the projects that have been proposed or are under construction have not been following appropriate procedures with regard to gaining clearance and undertaking consultation. They further suggest that in the building of some of these projects, parts of the Kishtwar High Altitude National Park might be submerged and that environmental mitigation measures, such as fish ladders, and benefit-sharing measures, such as adequate compensation for loss of land or employment, have not been given to local people. Further, many Indian-based civil society groups object to the number of projects proposed or under construction because no cumulative assessment has been carried out to determine the impacts. The Union Ministry of Environment and Forests (MoEF) has specifically requested that such cumulative assessments be carried out, but the government of Himachal Pradesh has asked for waivers on environmental clearance, suggesting that cumulative assessments are against the interests of the state. Such a perspective clearly disregards the ecosystem services and the livelihoods of people that are currently living in the Chenab subbasin and instead prioritize a model of development where large-scale projects can create economic growth and in doing so integrate economically poorer areas with wealthier ones, ostensibly to the benefit of both regions. Clearly, though, investment is in itself not enough; there must be a change in the relationship between local people and energy suppliers.

8.6 Imagining a New Management Across the Indus Between India and Pakistan

The preceding sections of the chapter have clearly indicated that the Indus basin is at an important juncture in terms of management and the possibilities for sustainable development. Development in the basin is all too often exclusionary and serves to inflame tensions between and within Pakistan and India, rather than being a catalyst for joint prosperity. Moreover, it is not only just in terms of human security that a new approach is demanded; the projected cumulative environmental effects of large-scale transformation of the Indus also clearly demand the most serious consideration.

Clearly, then, in order to increase accountability, transparency, and legitimacy of water resource planning in the Indus basin, there is a need to include a range of other stakeholders and to craft robust and durable institutional arrangements that ensure their participation in the design and implementation of projects. How can such an institutional transformation be achieved in the Indus? One promising trajectory of change may be encouraging dialogue through a range of what Dore (2007) calls multi-stakeholder platforms. There is certainly a central role for track 1 discussions between the governments of India and Pakistan as is currently occurring as part of the composite dialogue. However, beyond these formal bilateral processes, what is needed to build respect and capacity across the basin are a set of processes that encompass think tanks, parliamentarians, nongovernment organizations, and civil society. These can contribute to, and be engendered by, the de-securitization of the politics of water at a range of scales (Pohl et al. 2014).

Several authors have argued that there are hopeful developments occurring in the Indus and that these suggest new directions in the way water governance is approached (see Hill 2013a, 2015). The World Bank has thus far sponsored six rounds of the Abu Dhabi Dialogues under its South Asia Water Initiative (SAWI) and has now expanded to include a knowledge platform. These dialogues are non-formal and are intended to build a shared basis of understanding. While they include all seven countries involved in Himalayan water sharing, they nevertheless represent an opportunity for stakeholders from Pakistan and India to understand each other's perspectives, with the last iteration (2012) having a focus on transboundary collaboration for floods and disaster management, including in the Indus (International Bank for Reconstruction and Development 2013, p. 4). Moreover, the small grants component of SAWI has, among other things, a specific focus on the social dimensions of climate vulnerability in the Indus basin (International Bank for Reconstruction and Development 2013, p. 34).

As well as initiatives driven explicitly by multilateral organizations, there is a burgeoning corpus of reports that have resulted from dialogues on the Indus, usually organized jointly between Indian and Pakistani think tanks with collaboration from outside agencies such as the Atlantic Council or the Stimson Center (Indus Basin Working Group 2013; Ahmad 2012). These serve as an important guide or road map to the broader debates about how the Indus can be governed that seeks to look beyond the dominance of a supply-side hydraulic paradigm that remains the frame of reference of water bureaucracies and to incorporate a number of other factors, such as ecosystem services, adaptive management, and capacity building. The roundtables conducted as part of these processes are also important in building links across the Indus because they can go beyond the twisting and turning of Indo-Pakistan politics that frequently center on specific contentious projects.

An example of local think tanks and civil society trying to (among other things) create an alternative dialogue about water sharing in the region is an initiative called Imagine a New South Asia (INSA), which brought together groups from across South Asia. Originally begun with a secretariat housed at the Bangladesh Unnayan Parishad (BUP), in its most recent iteration, INSA was housed at the Sustainable Development Policy Institute (SDPI) in Islamabad. Indications are that INSA began strongly but has lost momentum after a few years, with the organizations involved not being able to spare time or resources for an initiative that is not project based and funded by donors (Hill 2013a, b; Action Aid 2011). These problems are not confined to INSA and are a familiar tale when considering many similar initiatives that struggle to sustain themselves across South Asia.

In a more general sense, multitrack diplomacy initiatives in the Indus basin are infrequently able to penetrate into government policy and offer significant criticism of government, lessening this potential to an even greater extent. Indeed, in a broader sense, civil society operates in extremely tight spaces with regard to being given "a seat at the table" in the region, particularly in Pakistan; and there is much greater space given to service delivery than there is to advocacy groups. Forging long-standing relationships between groups across the Indo-Pakistan national borders are extremely problematic. Moreover, those that are able to be involved in this way are often themselves part of a relatively narrow circle of bureaucratic elites; thus while their presence in these dialogues is important, there are likely to be significant gaps between these technocratic experts and many of the constituents they speak on behalf of. Two decades ago Dipak Gyawali criticized a similar NGO study, led by the Centre for Policy Research and funded by the Ford Foundation, on India's transboundary water relations with its eastern neighbors. There, Gyawali noted the tendency to call only upon ex-bureaucrats to participate in such studies. In his view, while they would bring experience and access to data, they were not ideal candidates to move toward alternative approaches and their input would usually only bring a "reiteration of the conventional past" (Gyawali 1995). On the other hand, those groups who are truly trying to challenge the institutional status quo are often excluded from such dialogues.

Perhaps the boldest new management proposal to come along is the proposal to demilitarize the Siachen Glacier, a high-altitude arena of conflict since 1984, and to create an International Peace Park or equivalent (Ali 2008). Attractive as it may appear, prospects for this proposal are exceedingly dim. Indeed, as Baghel and Nuesser (2015) have recently noted, the Indian army leadership has become unusually strident in its assertions of the need for a continuing military stronghold on the Siachen Glacier, which would seem to exacerbate the intractability of the standoff over this disputed territory.

8.7 Capacity Building, Education, and Facilitating Voices from Below

An important part of building sustainable institutions across the Indus is capacity building (Indus Basin Working Group 2013). The culture of technocratic bureaucracy in an institution like Pakistan's Water and Power Development Authority (WAPDA) has drawn from the strength of Pakistan's engineering education; while endorsing the intellectual traditions of the supply-side paradigm, it has also been a mechanism for mobility for young engineers and so has perpetuated itself as the dominant understanding of water resource management. However, there is a lack of depth of experienced professionals in a range of key and evolving areas, including things as diverse as climatology (Indus Basin Working Group 2013), ecosystem management, risk assessment, sustainable development, and impact assessment. Arguably the stakes in the Indus are too high and the issues too complex to confine the kinds of education to a rigid positivist model that privileges the knowledge of engineers and hydrologists while providing little legitimate voice for other forms of scientific and social scientific knowledge. Those other forms of knowledge will arguably be vital for ensuring a sustainable future for the Indus basin, and they will go a considerable way toward ensuring that biophysical and social factors are assessed together as part of the complex interface that exists between water and societies at a variety of scales.

This is a challenge that is not unique to South Asia, even if the dominance of the supply-side engineering-led paradigm is particularly acute in that region. Nevertheless, water education is evolving quickly with new programs being developed around the world. To some commentators, the need of the hour is to broaden the breadth of approaches covered to augment the traditional study of engineering with greater attention to the natural sciences. Others argue that more attention should be given to covering both natural sciences and social sciences. To some extent these debates reflect lingering disagreements about whether water management is a matter of resolving technical challenges, which may extend beyond engineering to encompass other natural sciences, or is rather predominantly a question of developing approaches that can comprehend the trade-offs required in juggling competing ideas around what constitutes development and different valuations of the environment. What is certainly true in the case of the Indus is that the current training does not adequately equip professionals to consider issues holistically, and it remains, as Gyawali put it with reference to other parts of South Asia, a "two-legged stool" that fails to understand the broader societal context of water resource development.

As well as contributing to a broader and deeper understanding of water's role in the societies of the Indus basin, perhaps the biggest role for water education in the region is in building future institutions that can respond to uncertainty, risk, and hazards. This is particularly important in the Indus basin given that projections of a changing climate suggest that these issues will become all the more pertinent in the future (Michel, this volume; Jaitly 2009). How do we build institutions that can be responsive to this? Arguably the prevailing institutional culture that is underwritten by technocratic certainties is ill suited to these challenges; this is not surprising, as consideration of risk is poorly integrated into most approaches to water governance, including university curricula (Read and Kuhl 2015): but it is also clear that capacity building and education have a significant role in preparing future water managers in the Indus basin.

8.8 Media, Capacity Building, and Its Role in the De-securitization of the Indus Basin

As well as increasing the breadth and depth of education and capacity building, there is also no doubt that the media is an important mechanism for increasing awareness and de-securitizing the Indus. Indeed, as has been demonstrated elsewhere (Hill 2013a), the media has a significant role in the way that the Indus problem is constructed and understood both within and outside the region. Certainly, censorship, both official and self-censorship, plays a role in how media on either side of the border reports water-sharing issues. More generally, in an increasing media-saturated world of 24/7 news channel cycles in both countries, it is often the case that all sorts of strongly held views are given airtime, arguably as a way of generating ratings. This difficulty is tied into the way that water is defined as a national security issue and has the effect of trying to make the issue predominately understood in adversarial terms as a bilateral issue framed in terms of India versus

Pakistan. Arguably this also has the effect of reducing the capacity to talk about the complex outcomes resulting from the modification of rivers and its impacts on people and the environment.

While little examined in academic circles, and receiving even less attention among policymakers, there are nevertheless some encouraging initiatives that have taken place that seek to change the representation of the India-Pakistan conflict, including over how the waters of the Indus are shared. In the past two decades, there has been a range of citizen diplomacy initiatives focused on what Galtung calls peace journalism (Faiz 2007). For example, Sherndurnikar (n.d.) has provided a content analysis of the Aman ki Asha (Hope for Peace) program that was launched by Jang Group (Pakistan) and *Times of India* (India) media houses in 2010. Through a range of different events (such as literary and cultural events and editor workshops) as well as a well-regarded television series (Aman ki Asha), this program has sought to draw attention to the commonalities of experience of people living on both sides of the India-Pakistan border. It has included a significant proportion of content devoted to water issues in the Indus basin.

Unfortunately, this series subsequently became embroiled in the larger fallout associated with the Pakistani establishment's actions against Geo TV and the Jang group, following allegation against the ISI and the Pakistan military by that channel (Sarwar 2014). Indeed, some critics, including the Chairman of the Pakistan Tehreek-i-Insaf (PTI), Imran Khan, have suggested that Aman ki Asha was unduly biased toward India and sought to portray Pakistan in a poor light (Shaukat 2014).[1] In this sense, the Aman ki Asha initiative demonstrates both the potential of media to create new lines of dialogue and the significant obstacles that such actions may face.

8.9 Conclusion

Any change in the Indus basin is constrained by a formidable combination of broader political relations, vested interests, and a technocratic approach to water resource management that has the effect of discursively delimiting the debate about what can and should be done. As such, the opening of new political spaces around Indus water governance is a long and unenviably difficult process, and there are no guarantees of any success in this regard. It is also clear that if water management is to have greater legitimacy among local stakeholders than is currently occurring, new approaches must be utilized; but to do so is to challenge a range of power structures that currently exist. These new approaches require multidisciplinary education and capacity building, media training, and greater space for nongovernment

[1] Geo's license for three of its television channels, Geo News, Geo Entertainment, and Geo Tez, was suspended for 15 days in May 2014 as a consequence of the allegations against the ISI. During the same period, leading Geo TV news anchor Hamid Mir was subject to assignation attempts. For a review of this controversy, see Sarwar (2014).

organizations. A key argument in this chapter has concerned the potential for economic partnerships to bring shared prosperity across the basin. However, it has also been argued that for such potential to be realized to the benefit of those living in the more economically and politically marginalized parts of the basin, this integration cannot simply replicate the previous experiences of hydropower development in the Indian- and Pakistani-held parts of the Himalayas.

As we have seen, in recent years a significant counter-discourse has emerged from stakeholders beyond the Indian and Pakistani states, which seeks to imagine Indus basin-wide management in ways that can augment and ultimately transform existing approaches (Indus Basin Working Group 2013; Bakshi and Trivedi 2011; ICPC n.d). These counter-discourses argue for a move beyond the supply-side hydraulic paradigm that remains the dominant frame of reference of water bureaucracies in South Asia. Basin-wide management requires these institutions to consider both upstream and downstream communities and in doing so pay equal attention to the watersheds, catchments, and headwaters of the Indus basin (Rasul 2010, 2014). However, translating these recommendations for enhanced attention to inclusive development, ecosystem services, and adaptive management into tangible actions requires capacity building as well as institutional transformation. This process is not simply a mindset shift for policymakers but is instead an inherently political process that challenges vested interests at a range of scales across the basin.

In seeking to document and analyze those initiatives that seek to build alternative institutions and approaches to water governance in the Indus basin, this chapter has argued that cooperation can take different forms and operate at different scales. Certainly, there is a range of initiatives that seek to increase the participation on non-state actors. On the other hand, while many of these show significant potential, the chapter has argued that there are constraints to the extent to which they can influence policymakers at the present time. As such, while the future of the Indus basin remains an open question, it would seem most likely that equitable and sustainable management across the basin will remain a distant and optimistic endeavor that is unlikely to be fully realized without much more significant political changes within and between India and Pakistan.

References

Action Aid International Asia (2011) Natural resource management in South Asia: imagine a New South Asia. Pearson Education, New Delhi

Ahmad S (2012) Water insecurity: a threat for Pakistan and India. Atlantic Council of the United States, Washington, DC

Ali SH (2008) The Siachen Peace Park proposal: moving from concept to reality. Environment 50(3):43–43

Ali I (2004) Historical impact on political economy of Pakistan. Asian J Manag Cases 1(2):129–146

Asian Development Bank Institute and Asian Development Bank (2013) Connecting South Asia and Southeast Asia: interim report. Asian Development Institute, Tokyo

Baghel R, Nüsser M (2015) Securing the heights: the vertical dimension of the Siachen conflict between India and Pakistan in the Eastern Karakoram. Polit Geogr 48:24–36

Bakshi, G. and S. Trivedi (2011) The Indus equation. Strategic Foresight Group, Mumbai

Briscoe J (2010) Troubled waters: can a bridge be built over the Indus? Economic and Political Weekly, pp 11–17

Briscoe J, Qamar U (2006) Pakistan's water economy: running dry – background papers. The World Bank, Washington, DC

Chellaney B (2011) Water: Asia's new battleground. Georgetown University Press, Washington, DC

Committee on Foreign Relations (2011) Avoiding water wars: water scarcity and central Asia's growing importance for stability in Afghanistan and Pakistan. A majority staff report one hundred twelfth congress, first session 2011. U.S. Government Printing Office, Washington, DC

D'Souza R (2008) Framing India's hydraulic crisis: the politics of the modern large dam. Mon Rev 60(3):112–124

D'Souza R (2014) Peace is not possible and war is not an option! Should we still be 'talking up' nontraditional security? Strateg Anal 38(5):741–748

Dore J (2007) Multi-stakeholder platforms: unfulfilled potential. In: Lebel L, Daniel R, Koma YS (eds) Democratising water governance in the Mekong region. Mekong Press, Thailand, pp 197–226

Faiz A-u-H (2007) India-Pakistan dialogue: bringing the society in. Regional Center for Strategic Studies, Colombo

FAO (2008) Evaluation of FAO cooperation in India: Andhra Pradesh farmer managed groundwater systems evaluation report. Food and Agriculture Organization of the United Nations, Rome

Gazdar H (2005) Baglihar and politics of water: a historical perspective from Pakistan. Econ Polit Wkly 40(9):813–817

Glassman J (2010) Bounding the Mekong: the Asian development bank. University of Hawai'I Press, Honululu

Gyawali D (1995) Flood of hope; a review of the output of the techno-exuberant school of thought. Himal Association, Lalitpur

Hill D (2003) Food security, governance and rural development in India under the BJP. South Asia J South Asian Stud 25(3):147–164

Hill DP (2008) The regional politics of water sharing: contemporary issues in South Asia. In: Lahiri-Dutt K, Wasson R (eds) Water first: issues and challenges for nations and communities. Sage, New Delhi, pp 59–80

Hill DP (2009) Boundaries, scale and power in South Asia. In: Devleena G, Heather G, Stephanie D (eds) Water, sovereignty, and borders in Asia and Oceania. Routledge, New York, pp 87–103

Hill DP (2013a) Trans-boundary water resources, crisis and uneven development in South Asia. South Asia J South Asian Stud 36(2):243–257

Hill DP (2013b) Water-sharing in the Indus basin: a peaceful, sustainable future is possible. Glob Dialogue 15(2). http://www.worlddialogue.org/content.php?id=558

Hill DP (2015) Where hawks dwell on water and bankers build power poles: transboundary waters, environmental security and the frontiers of neo-liberalism. Strateg Anal 39(6):729–743

Homer-Dixon TF (1994) Environmental scarcities and violent conflict: evidence from cases. Int Secur 1(5):5–40

Homer-Dixon TF (1999) Environment scarcity and violence. Princeton University Press, Princeton

Indus Basin Working Group (2013) Connecting the drops: an Indus basin roadmap for cross-border water research, data sharing and policy coordination. Stimson, ORF and SDPI, Washington, DC

International Bank for Reconstruction and Development (2013) South Asia water initiative 2009–2013 final report. World Bank, Washington, DC

Jaitly A (2009) South Asian perspectives on climate change and water policy. In: Michael D, Pandya A (eds) Troubled waters: climate change, hydropolitics, and transboundary resources. The Henry L. Stimson Center, Washington, DC, pp 17–31
Lahiri-Dutt K (2000) Imagining rivers. Econ Polit Wkly 35(27):2395–2397, 2399–2400
Lahiri-Dutt K, Wasson RJ (eds) (2008) Water first: issues and challenges for nations and communities in South Asia. Sage, New Delhi
Lieven A (2012) Pakistan: a hard country. Penguin, London
Mazoomdar J (2014) Invisible strings- the IB's report on NGOs'. Tehelka 26 (11), 28th June 2014
McCartney MP, Sullivan C, Acreman MC (2001) Ecosystem impacts of large dams' background paper No. 2 prepared for IUCN/UNEP/WCD (International Union for Conservation of Nature and Natural Resources and the United Nations Environmental Programme)
Molden DJ, Vaidya RA, Shrestha AB, Rasul G, Shrestha MS (2014) Water infrastructure for the Hindu Kush Himalayas. Int J Water Resour Dev 30(1):60–77
Mollinga PP (2008) Water, politics and development: framing a political sociology of water resources management. Water Altern 1(1):7–23
Mustafa D (2010) Hydropolitics in Pakistan's Indus basin. U S Inst Peace. Special Report No. 261
Mustafa D, Akhter M, Nasralla N (2013) Understanding Pakistan's water-security nexus. United States Institute of Peace, Washington, DC
Pohl B et al (2014) The rise of hydro-diplomacy. Adelphi, Berlin
Pomeranz K (2009) The great Himalayan watershed: agrarian crisis, mega-dams and the environment. New Left Rev 58:5–39
Pradhan NS, Khadgi VR, Schipper L, Kaur N, Geoghegan T (2012) Role of policy and institutions in local adaptation to climate change – case studies on responses to too much and too little water in the Hindu Kush Himalayas Kathmandu: ICIMOD
Rasul G (2010) The role of the Himalayan mountain systems in food security and agricultural sustainability in south Asia. Int J Rural Manag 6(1):95–116
Rasul G (2014) Food, water and energy security in South Asia: a nexus perspective from the Hindu Kush Himalayan region. Environ Sci Pol 39:35–48
Read L, Kuhl L (2015) Bringing the elephant into the room: integrating risk into interdisciplinary water programmes. J Contemp Water Res Educ 155:19–27
Rodell M, Velicogna I, Famiglietti JS (2009) Satellite-based estimates of groundwater depletion in India. Nature 460:999–1002
Roy A (1999) The Cost of Living. Flamingo [imprint HarperCollins], London
Salman MAS, Uprety K (2002) Conflicts and cooperation on South Asia's international rivers; a legal perspective. The World Bank, Washington, DC
Sarwar B (2014) Pakistan's media wars. Himal South Asia, July. The South Asian Trust, Lalitpur
Shaukat S (2014) The myth of 'Aman Ki Asha' South Asia media.net. http://www.southasianmedia.net/sam-monitors/pakistan-media-crisis/the-myth-of-aman-ki-asha-story
Sherndurnikar N (n.d) Content analysis of 'Aman Ki Asha' a peace journalism campaign by the times of India and Jang Group Pakistan'. Available at https://www.academia.edu/3237852/CONTENT_ANALYSIS_OF_AMAN_KI_ASHA_A_PEACE_JOURNALISM_CAMPAIGN_BY_THE_TIMES_OF_INDIA_AND_JANG_GROUP_PAKISTAN
Wilson JS, Otsuki T (2007) Regional integration in South Asia: what role for trade facilitation? vol 4423, World Bank Policy Research Working Paper. World Bank, Washington, DC, http://elibrary.worldbank.org/doi/pdf/10.1596/1813-9450-4423
World Bank (2014) Regional integration in South Asia March 24th, 2014 www.worldbank.org/en/region/sar/brief/south-asia-regional-integration. Accessed 24th Aug 2015

Newspaper Articles

'India for resolving Ratle hydro plant issue with Pakistan through talks' Economic Times 23 Aug 2015. http://economictimes.indiatimes.com/industry/energy/power/india-for-resolving-ratle-hydro-plant-issue-with-pakistan-through-talks/articleshow/48638231.cms. Accessed 30 Aug 2015

India not causing water shortage in Pakistan: Irsa', The News International July 12, 2015. http://www.thenews.com.pk/print/13570-india-not-causing-water-shortage-in-pakistan-irsa [last accessed 16th June, 2016]

Alam I (2015) Pakistan to move ICJ over India's water aggression. The Nation, 9 Feb 2015

'PDP Perturbed over Power minister's rejection of NHPC projects', Early Times Report 5/22/2016. http://www.earlytimes.in/newsdet.aspx?q=176588 [last accessed 16th June, 2016]

Shah U (2015) Time to call return of power projects from NHPC. Kashmir Monitor, 4 Mar 2015

Water treaty with Pakistan not profitable, says Omar Abdullah'. The Asian Age, 24 Aug 2014

Chapter 9
The Indus Basin: The Potential for Basin-Wide Management Between China and Its Himalayan Neighbours India and Pakistan

Srikanth Kondapalli

Abstract Despite the presence of huge water resources, several factors are contributing to water security issues in Asia in general and southern Asia in particular. From this region flow some of the major rivers in Asia, such as Yangtze, Yellow, Indus, Yarlung Zangbo/Brahmaputra, Salween, Mekong, Irrawaddy and others. These rivers drain several million square kilometres and have become lifelines for the food security of billions, apart from transporting goods and services and for industrial development. Countries such as China, India and Pakistan sit on enormous water reserves in this part of the world, and these are in the recent period triggering securitisation of water-related issues due to a number of reasons. China has the fourth largest freshwater reserves in the world. However, due to increasing demands over water use, such resources are being increasingly and extensively exploited for economic purposes. These issues have triggered wide debate among officials, lawmakers, scholars, environmentalists and others. In order to address water scarcity issues, China recently launched several initiatives and programmes such as the South-to-North Water Diversion Project (SNWDP), construction of either water diversion dams or hydropower dams and the like. It is argued in this chapter that while in the overall water discourse of China, the Indus River takes a marginal seat in comparison with other major rivers, China has followed a two-pronged approach, namely, stop-gap understandings on water sharing or, more accurately, water measuring with the immediate lower riparian states, including India, while actively exploring coop erative efforts with the lowest riparian state, Pakistan, in regard to dam construction and hydroelectricity generation, including even eventual protection of these facilities with China's paramilitary/military forces in the longer run. Thus, cooperative efforts do exist in the Indus basin between China and Pakistan, while in the case of China and India, both cooperative and competitive elements are forthcoming.

Keywords Indus basin • Water sharing • China-Pakistan cooperation • China-India competition

S. Kondapalli (✉)
Chinese Studies, Jawaharlal Nehru University, New Delhi, India
e-mail: srikondapalli@gmail.com

Z. Adeel, R.G. Wirsing (eds.), *Imagining Industan*, Water Security in a New World, DOI 10.1007/978-3-319-32845-4_9

9.1 China's River Systems

China is endowed with rich water resources originating mainly from the western portion of the country. It has an estimated 50,000 rivers of small, medium and large systems (China's Ministry of Water Resources website). The following table provides a list of main rivers in China.

China's Rivers

	Length (km)		Catchment area (km²)		Average annual surface runoff (billion m³)
Name	Total	In China	Total	In China	
Yangtze River	6,300		1,808,500		975.5
Yellow River	5,464		752,000		59.2
Pearl River	2,214		454,000	44.2	336
Haihe River	1,090		263,400		22.8
Huaihe River	1,000		269,000		61.1
			187,000		44.3
Songhuajiang River	2,309		556,800		74.2
Liaohe River	1,345		219,000		14.8
Heilongjiang River	4,440		1,855,000	89.11	355
Lancangjiang River	4,500	1,612	810,000	15.4	74
Erqisi River	4,248	633	1,643,000	5.73	10
Nujiang River	3,200	1,659	325,000	13.78	68.9
Yarlung Zangbo River	2,900 (916 in India)	2,057	935,000	24.0	165.4
Tarim River	2,300	2,046	198,000	19.4	20.5
Yili River	1,500	601	131,000	6.16	17
Yuanjiang River	1,183	565	158,000	7.63	18.3
Wusuli River	890		187,000	5.67	45.
Yalujiang River	795		63,800	3.25	29.1
Tumenjiang River	520		300,000	2.2	7.52

Source: China's Ministry of Water Resources and Li Zhifei (2015: 73)

In the northwest and southwest regions of China, specifically in Tibet, are located a number of important river systems (Zhifei 2015, 66). The Yarlung Zangbo and Nujiang rivers have a catchment area of 624,000 km² in China. With nearly 2,900 km of journey and dropping from 13,000 ft at its origins to about 5,000 ft at Namcha Barwa at the disputed border between Tibet and Arunachal Pradesh, the Yarlung Zangbo/Brahmaputra feeds more than 100 million people in Tibet, India and Bangladesh (Sud 2008). In 2006, China's first national water resources survey concluded that China has a potential of 500 million kW of hydroelectricity generation capacity of which only 117 million kW (24 %) was realised by 2005. The report suggested that greater unused potential capacity is exhibited in water resources of south-western China in the Jinsha, Yarlung Zangbo, Hongshui and Lancang rivers (National Development and Reform Commission 2006).

These river systems sustain a huge number of people. They are also termed as "sorrows" for the people for the fury they unleash in the form of periodic floods and other natural disasters. Taming these rivers reached mythical proportions in the history of China, memory of which still lingers in contemporary political debates (Hong and Shengqi 1998). While Generalissimo Chiang Kai-shek utilised water as a potent weapon to unleash damage on the invading Japanese forces in World War II, Mao Zedong and others of the "first generation" of communist leadership controlled the rivers through extensive dam construction activities beginning from the 1950s. Subsequent generations of leadership in China furthered this process, although in the latest phase, certain debates were allowed to come out into the public sphere, partly due to the aggravating environmental situation. Among all these "generations", securitisation of water issues has been a strong phenomenon with "mainstream" views contending that the river waters need to be tamed further through different projects.

Three main strands are visible in current Chinese assessments of water security issues. One, there is a need to exploit water resources of the country for economic development, especially in view of Deng Xiaoping's motive of getting rich quickly. A second and recent strand, in the light of the devastating environmental fallout of China's gigantic hydroelectric, storage reservoir and canal projects, is that there should be "coordinated" development between taming rivers and ecological protection. Nevertheless, the concern is again mainly on "sustainable" economic development (Zhang 2004). A third – connected to the international dimension – relates to China's having entered into a memorandum of understanding (MoU) with a number of lower riparian states focused on hydrological data exchange without any long-term binding provisions impacting China (Zhifei 2015).[1]

[1] In August 2002, China revised the 1988 water-related legislation into a law. It stated that the state owns the water resources of the country. In the supplementary provisions (article 78), this law suggested that the international understanding of China on water issues supersedes the domestic law (Water Law of the PRC 2002). As a part of border dispute resolution with its neighbours, China has also considered the transboundary river waters, but many of these are in the domain of management issues. For instance, with Bangladesh, an MoU was signed for hydrological information in 2008; with India in 2005, 2008, 2010 and 2015 for hydrological information; water quality on Irtysh River in 2011; joint diversion on Horgos in 2010; exchange of hydrological information and disaster prevention with Kazakhstan; commercial navigation in 2000, transportation in 1994 and law enforcement on Mekong in 2011; management with Russia in 1962, 1988, 1994 and 2010; and so on (Chen et al. 2013; Zhifei 2015, chapters 3 and 5; China International Water Law).

9.2 South-to-North Water Diversion Project (SNWDP)

In order to utilise its water resources, China has undertaken a number of initiatives, including the recently launched South-to-North Water Diversion Project (SNWDP) [*Nanshui beidiao*] which was proposed in 1952 by Mao Zedong. He reportedly stated: "The south has a lot of water, the north little. If possible, it is ok to lend a little water" (Mao cited in Nickum). After half a century of debate on the project, the State Council finally decided to go ahead with the construction of the project in 2002 (Liu 1998). In December 2002, work began on the eastern route to divert waters to Shandong province. The central route work began in December 2003 and was completed by 2010. It supplies waters to Henan, Hebei, Beijing and Tianjin. The SNWDP is projected to divert 44.8 billion cubic metres of water annually from the Yangtze River through the three routes by 2050. The project also plans to link up four major river systems of China, namely, the Yellow, Yangtze, Huaihe and Haihe rivers. The following table provides information on the SNWDP:

Route	Eventual diverted volume	Diversion extent	Timeline	Estimated costs		
				Stage I 2000–2010	Stage II 2010–2020	Stage III 2020–2050
Eastern route	14.8 billion m^3/year	1,156 km	December 2002–2006	17.9 billion yuan	11.3 billion yuan	–
Central route	13.0 billion m^3/year	1,267 km	December 2003–2008	23.4 billion yuan	31.5 billion yuan	–
Western route	17.0 billion m^3/year	1,300 km	2010–2050	–	0–20 billion yuan	230–250 billion
Total	44.8 billion m^3/year	3,723 km	–	41.3 billion yuan	42.8–62.8 billion yuan	230–250 billion yuan

Source: "South-to-North Water Diversion Project, China" accessed from http://www.water-technology.net/projects/south_north/ and James E. Nickum

The first two routes were expected to cost about $26 billion, with the middle route costing about $18 billion. The State Council earmarked $37.2 billion for phase I of the project in eastern and middle routes. (Xinhua 2008) By 2008, sections of the project in Shandong and Jiangsu provinces along the eastern route had been completed, apart from the 210 km section linking Shijiazhuang with Beijing (AFP 2008). In December 2007, China began digging a tunnel 7.8 km in length beneath the Yellow River in Shandong province to divert 442 million cubic metres of water from the Yangtze to the Yellow River. This was expected to cost about $92.1 million (Xinhua, December 28, 2007). By the end of 2015, China's Ministry of Water Resources estimated the cost of the completed project at 500 billion yuan ($82 billion) (China's Ministry of Water Resources 2015).

The western route is the largest, most difficult and costliest of all the routes. Expected to cost about $36 billion, this route is to divert 17 billion cubic metres of water to the Yellow River by 2050. This route has generated more controversy than the earlier two because of the large scale of diversion of the waters as well as the potential international dimension of the project. That is, lower riparian states such as India and Bangladesh have concerns about the proposed diversion of waters as well as about the construction of a hydroelectric dam or dams anywhere on the Brahmaputra, but especially any contemplated near the river's so-called Great Bend. Such plans were initially denied by the Water Resources Minister Wang Shucheng and by a foreign ministry spokesman (Reuters 2006). Nevertheless, in 2003 it was reported that Chinese scientists had conducted a feasibility study for a hydroelectric project on the upper reaches of the Brahmaputra mainstream (Hodum 2007). The project – the 510 MW Zangmu hydropower dam – began in 2008 and became operational in 2015.

While China has utilised these river systems extensively for agricultural, hydroelectrical or transport purposes, the lower riparian states are also heavily dependent on these waters for similar purposes. This issue is lately coming to the fore, some even suggesting the potential for conflict that this issue could generate in the twenty-first century between China and these states (Chellaney 2011; Hodum 2007). Although China has been reluctant to acknowledge the "transboundary" nature of these river systems, it is compelled to come to an understanding with several lower riparian states in the light of increasing economic interdependencies. Rivers flowing from China through Kazakhstan, India, Pakistan, Bangladesh, Myanmar and the Indochina region are increasingly giving rise to contentious issues owing to their significant impact on the ecosystem and livelihood patterns of hundreds of millions of people downstream. If this transboundary issue is not addressed in an amicable manner, there are several chances for conflict to emerge in regard to river water-sharing arrangements (Economy 2004, 2008; Menniken 2007; Zhifei 2015: 136).

9.3 The Indus Basin

The focus of the above discussion is mainly on China's major rivers, including the recent controversy over the Yarlung Zangbo/Brahmaputra River, with marginal focus on the Indus. The Indus River originates in Tibet at Lake Manasarovar in the Mount Kailash region and flows westwards through India and Pakistan, with several tributaries (including the Kabul River from Afghanistan) and small rivulets joining on the way, before it enters the Arabian Sea (Central Tibetan Administration 2000). In the course of its run, the Indus has shaped an entire civilisation in its tracks – the Indus Valley Civilisation (Mountjoy 2005). Alexander's expeditions in the region were to have a lasting impact on subsequent ages (Huang 2010).

The Indus River basin is over one million square kilometres in size with a major portion of the river flowing in Pakistan (47 %), the rest in India (39 %), China (10 %) and Afghanistan (6.6 %) (the Indus Waters Treaty 1960). Due to the unresolved territorial dispute between China and India, the extent of the Indus River basin is contested. The 1960 Indus Waters Treaty between India and Pakistan regulated water usage between the two; however, contention over this usage has in recent years led to some acrimonious disputes (Zhong et al. 2011; Hu et al. 2010; Chengdu

Business Report 2010). There are also new factors that have contributed to tensions, including population growth, rapid urbanisation and industrialisation, environmental degradation due to deforestation and inefficient use of resources (Yunhui et al. 2011).

According to a United Nations report, the annual flow from China to India in the Indus basin is 181.62 km^3 and within India is 50.86 km^3, resulting in a flow from India to Pakistan in this part of 232.48 km^3, of which 170.27 km^3 are reserved for Pakistan and 62.21 km^3 are available for India (United Nations Food and Agriculture Organization; Wolf et al. 1999). No treaty exists between China and India on sharing water resources, although some Chinese have argued that the Indus Waters Treaty could be the basis for such treaties with the Southeast Asian countries (Bai 2012). The main threats to the Indus River, according to a *People's Daily* commentary comparing ten major rivers, are climate change, overutilisation of water, contamination of water and dam construction activities (*People's Daily* 2007).

The Indus River

River/tributary	Length (km)	Catchment area (km^2)	Remarks
Indus	2,900 (1,114 km in India)	372,000	Originates near Mount Kailash
Chip Chap River in Aksai Chin/Karakoram ranges		4,410	Disputed between India and China. Enters the Shyok River and then Indus
Sutlej River			Originates near Rakshastal in Tibet and enters at Shipki La in India
Galwan in Aksai Chin			Disputed between India and China. Joins the Shyok River and then Indus
Shiquanhe River			

Tributaries of the Indus River

River	River length (km)	Drainage area (10,000 sq km)	Origins	River flows
Jhelum	774	6.3 5	Kashmir, India	Chenab
Chenab	1,200	13. 80 (incl Jhelum)	India	Dry
Ravi	725	1.16	India	Chenab
Beas	470		India	Sutlej
Sutlej	1,450	39.5	Tibet, China	–
Shyok	550		Tibet, China	Dry
Kabul	700	8.5	Afghanistan	Dry
Gumal	240		Pakistan	Dry

Source: Zhong Huaping et al. (2011: 68)

Water Resources Characteristics of the Indus River Basin

Country	Length (km)	Basin area (10,000 km²)	%	Runoff BCM	%	Main rivers involved
India	450	38.16	33.5	1,435	69	Indus and its five east tributaries
Pakistan	2,300	59.77	52.5	395	19	Indus and two on the west side, four tributaries on the east side
Afghanistan	–	7.21	6.3	–	–	Two tributaries on the west side
China	430	8.74	7.7	250	12	Sutlej and the east side of the tributary
Total	3,180	113.88	100	2,080	100	–

Source: Hu Wenjun et al. (2010: 1919)

China and Pakistan have entered into cooperative agreements in the last decade or so to utilise water resources for agricultural purposes and for generating hydro-electricity. A number of contracts were signed by both to construct such projects as the following table indicates. A number of Chinese engineers and workers have been operating at these projects (*The Nation* August 2, 2010).

Indus Water Utilisation and Developments

Dam	Details	China's role
Bunji Dam in Gilgit-Baltistan	190 m high; 7,100 MW capacity	China Three Gorges Project Corporation signed an MoU in August 2009 for this largest hydropower project in Pakistan
Chashma Barrage in 1971	11 m high; 184 MW – operational	56 km downstream of Chashma Barrage is the Jinnah Hydropower Project being built with Chinese company Dongfang Electric Corporation at $128 million
Allai Khwar river project in Besham District from 2012	61 m high; 560 GWh/year	Guangdong Yuantian Engineering Co. involved in the construction
Khan Khwar at Besham	46 m high; 306 GWh/year	China's Dongfang and Sinohydro involved
Gomal Zam	17 MW	China National Water Resources and Hydropower Engineering, Harbin Power Engineering Co. and Synohydro involved in the construction
Jabban power station, Malakand	22 mw	Chaozhou Huineng Electric Machinery, Zhejiang Jinlun Electromechanic Co. and Sanbian Sci-Tech Co.
Malakand-III in 2008		Harbin Electric Co. involved
Mangla Dam	146 m; 1,000 MW	China International Water & Electric Corp. involved in the raising of the height of the dam since 2004
Tarbela Dam in 1976	148 m high; 3,478 MW	China's Sinohydro is involved since 2013 in the Tarbela-4 Project for 1,410 MW at a cost of $928 million

Source: Pakistan Water and Power Development Authority and others

With the recent launch of the Silk Road initiative in 2013 and the visit of President Xi Jinping to Islamabad in April 2015, when China announced a massive $46 billion plan for investments in Pakistan, attention is now focused on hydroelectric projects across several rivers. Indeed, more than half of the $46 billion is to be spent on energy projects, specifically hydroelectric projects, in addition to infrastructure projects. Most of these will impact the disputed area of Kashmir. Hence, China appears to be upping the ante in its bilateral relations with India. Indeed, as is evident in the table above, China has been investing in infrastructure projects for some time in Pakistan-controlled portions of the disputed territory of Kashmir to the chagrin of New Delhi. Some of the emerging details of these investments appearing in the local press include the following:

- Diamer-Bhasha Dam on the Indus in the disputed Gilgit-Baltistan area for $12.6 billion – contract awarded to China's Three Gorges Project Corporation (Haider and Pearson 2015).
- CGGC-CMEC Consortium China is involved in the $1.5 billion Neelum-Jhelum Hydroelectric Power Project in Pakistan Occupied Kashmir which aims to divert the water of the Neelum (Kishanganga) River through a tunnel into the Jhelum River.
- Dasu Hydroelectric Project for $7.8 billion.
- Phandar Hydroelectric Project for $70 million.
- Bashu Hydroelectric Project for $40.01 million.
- Harpo Hydroelectric Project for $44.608 million.
- Yulbo Hydroelectric Project for $6 billion.
- China International Water & Electric Corporation has raised the level of the Mangla Dam in southern Mirpur district of Pakistan Occupied Kashmir by about 60 ft.
- China International Water & Electric Corporation for $2.1 billion Kohala Power Project at Muzaffarabad in Pakistan Occupied Kashmir.

Since the late 1980s, China's companies have begun to cooperate with Pakistan's river water projects. The Chinese Ministry of Water Resources' 13th Engineering Bureau, Guangxi International Economic and Technical Cooperation Company, the China Harbour Engineering Company, Three Gorges Project Corporation, Dongfang Electric Corporation, Guangdong Yuantian Engineering Co., Chaozhou Huineng Electric Machinery, Zhejiang Jinlun Electromechanic Co., Sanbian Sci-Tech Company and others have made major inroads into Pakistan's hydro projects. Most of these are state-owned enterprises and thus provide political leverage to China in the "all-weather" friendship between the two countries. However, China's entry into the Pakistan-controlled river basins is not without its problems. Apart from the security of its engineers and workers, who may be exposed to militant violence, there are also the tough conditions in the bidding process itself. According to Huang Lei, the Chinese companies have to compete with other international companies, specifically with the European and American companies preferred in Pakistan. Besides, the stringent quality requirements and other conditions are generally seen as posing constraints upon the Chinese bidder (Huang 1999). Another problem is that of huge sediment deposition, specifically in the Diamer-Bhasha Dam area (Tate et al. 2001).

While China has thus been cooperating extensively with Pakistan on Indus basin projects, its responses towards India are different. To a large extent, these responses were conditioned by Cold War logic and the balance-of-power approach, although there has been some softening in recent times, specifically in the signing of memoranda of understanding, if not in actual water-sharing treaties (Zhifei 2011). In relation to India, three recent trends suggest that water-related issues could be potential problem areas between these two countries[2]. As has been pointed out in the above tables, apart from the Indus River, there are also several tributaries and rivulets which merge with the Indus as it meanders towards the north and the west. In relation to India and China, these include the Chip Chap, Galwan, Shiquanhe and Sutlej rivers, with all of them disputed, whether over sharing, any artificial structures built on them impacting the downstream environment, or socio-economic impacts (Geng 2012; IDSA 2010). Due to the construction activity and the attendant deforestation in Tibet and Xinjiang, the water levels are falling and silt deposits are increasing, raising concerns downstream (Zhong et al. 2011: 154). In June 2005, an artificial lake on the upper reaches of the Sutlej River in Tibet burst leading to concerns on the lower reaches of Kinnaur district in Himachal Pradesh (Vinayak 2005). A year earlier, such flash floods had wrought havoc downstream (The Hindu, August 18, 2004). Unlike in 2000, when a similar outburst flood washed off apple cultivation in Himachal Pradesh, this time around China alerted the Indian side, although a visit by a four-member team from India to the site was put off (Arpi 2004). Also, the Depsang Plains incident between April 15 and May 6, 2013, in the western sector of the border between China and India, when the Chinese troops "intruded 19 km" and pitched tents inside the Indian claimed areas at Daulat Beg Oldi, had once again brought the river water dispute into the limelight with a blogger in China arguing that the area falls within China's sovereignty claim (Sina.com, April 28, 2013). River water issues thus have a clear bearing on the territorial dispute between the two (Wang 2013:11).

Under pressure from New Delhi, China has expressed its willingness to enter into an understanding with India on water issues, although the process is extremely slow and complicated. Both countries signed the first MoU on water issues during the visit of Premier Zhu Rongji to New Delhi in January 2002 for hydrological data exchange on the Brahmaputra (Ministry of Water Resources 2013–2014: 60). This was renewed later in 2008. In 2006, an expert-level mechanism was established during the visit of President Hu Jintao to India. This meeting format, repeated eight

[2] According to the Indian Central Water Commission report of 2015, up to the Indian border, the average annual water resources potential (billion cubic metres) of the Indus River is 73.3 (as compared to 537.2 for the Brahmaputra River); and in 2010 there was an estimated population of 57 million people living in the Indian-administered regions served by the Indus waters. These population figures are expected to increase to 69.2 million in 2025, to 81 million in 2050. Consequently, the estimated per capita average annual water availability (thousand cubic metres) is estimated to decline from 1,270 in 2010 to 1,059 in 2025 and 900 in 2050 (Indian Central Water Commission 2015: 31).

times between 2007 and 2014, discussed transboundary water issues. After Indian Prime Minister Manmohan Singh raised the matter of Brahmaputra River diversion projects at the Durban meeting with President Xi Jinping in May 2013, an MoU to renew the hydrological data exchange agreement was signed during Premier Li Keqiang's visit to Delhi in the same month. Further mention of this issue was made during Prime Minister Modi's visit to China in mid-2015; and during Vice President Li Yuanchao's visit to Delhi in late 2015, another MoU was signed. While the above were mainly addressed to the Brahmaputra River, in 2005 an MoU was signed dealing with hydrological data on the Sutlej River waters and this was renewed in 2010 (Ministry of Water Resources).

9.4 Conclusions

Thus while China undertook several major measures domestically in recent decades to alleviate water stress levels, its actions in regard to the international dimension of the rivers had created mixed prospects, with cooperation with Pakistan growing while tensions expanded with its other neighbours (Zhifei 2015; Chellaney 2011; Elhance 1999; Gleick 1993). The lower riparian states, including Kazakhstan, India, Pakistan, Bangladesh, Myanmar, Laos, Cambodia, Thailand, Vietnam and others, have become attentive towards China's water usage and control policies, although the latter has several non-binding MoUs signed with some of them.

Officially, the Chinese leadership had indicated that China would like to coexist peacefully with its neighbours to create a "harmonious world". China has initiated several policies aimed at furthering not only its comprehensive national strength (with diverted water resources partly factored in) but also revived "good neighbourliness" policies. Indeed, at the all-powerful 16th Communist Party Congress (2002), 17th Party Congress (2007) and 18th Party Congress (2012), relations with the neighbouring and developing countries had been given high priority. Nevertheless, several events in the international arena as well as at the regional level indicated that on water-sharing issues, China's responses have been largely bilateral in form but unilateral in content, with hardly any multilateral cooperative effort at all with the neighbouring and lower riparian states. A major exception is China's role in the Mekong Commission. Even here, however, the concerned Southeast Asian countries have recently voiced apprehensions on the environmental fallout of big dam construction in the upper reaches of the Mekong (Osborne 2015; Zaffos 2014). China's position on water issues could become problematic from the point of view of its soft power status in the immediate neighbourhood, not to mention the legal wrangling that might arise if the concerned lower riparian states were to resort to international jurisprudence and arbitration procedures as did the Philippines in regard to the South China Sea dispute. Another aspect is that China's actions on water issues could also lead to a balance-of-power approach by the affected lower riparian states. Although China has observer status in the South Asian Association for Regional Cooperation and has expressed its interest in becoming a full-fledged member, it is not clear so far whether the issue of water sharing between all these concerned states will be addressed in that forum or whether any coordination between the lower riparian states could take place there. Of course, China has been silent on this issue so far, confining itself to making arrangements at the bilateral level with Pakistan and Bangladesh. A major dilemma for China is that, on the path to its rise in the global economic and strategic matrix, water-sharing issues could pose serious problems for it. Hence, Beijing is exploring several options.

To start with, in the north-western regions of China flow important rivers that are transboundary in nature. One is the Irtysh River, which originates in the Chinese-held Altai Mountains and flows through Xinjiang, on into Kazakhstan's Lake Zaysan and finally into the Ob River in Russia through Omsk city. Another river, the Ili, drains water into Kazakhstan's Lake Balkhash after entering through Almaty city. China plans to divert these two major rivers, in addition to nearly 23 other smaller rivers and tributaries. This has raised concerns in Kazakhstan, as its industrial regions such as Karaganda and Pavlodar depend on the Irtysh. In addition, Kazakhs accuse China of seizing 150 mile^2 of Kazakh territory in 2001 for the control of the Black Irtysh River watershed. This is despite the border treaties between the two signed in 1996 and 1997. It has been reported that China has been unwilling to expand negotiations on water-sharing issues with the Central Asian states and Russia in the multilateral forum of the Shanghai Cooperation Organisation and instead treats this subject in a bilateral fashion (Gulati 2014). China (and Afghanistan) also refused to be part of the Central Asian initiative in Interstate Coordination Water Commission with prospects for conflict among these countries increasing (Castelein 2002: 114).

Secondly, at the international level, China, along with Turkey and Burundi, as upstream countries, opposed and voted against the 103 countries that supported the 1997 United Nations "Convention on the Law of the Non-Navigational Uses of International Watercourses". China refused to accept the words "transboundary" or "international" for waters flowing from one country to another (McCaffrey et al. in Menniken 2007: 102; Castelein 2002: 122). Further, China has also withdrawn its commissioner from participating in the World Commission on Dams meeting in 2000 (Menniken 2007: 102). These indicate the unilateralist position of China as an upstream country, and they also imply that China's dam construction activities in Tibet, Yunnan and other bordering provinces would not be governed or interrupted by international legal principles.

Thirdly, China's proposed plan for the construction of a dam at the Great Bend on the Yarlung Zangbo/Brahmaputra River as a part of the SNWDP is creating ripples in the lower riparian states of India and Bangladesh (McGormack 2001). The lower riparian states' concerns include the possible drying up of the delta region (due to silt deposition and water diversion), natural calamities (as the project traverses through earthquake-prone regions), harmful ecological changes and the like. As an experience in the Pearl River Delta in South China indicated, trapped silt in the reservoirs upstream led to the depletion of the deltas and lower reaches and/or decrease in fisheries and fertility downstream. While this benefited the farming communities at the reservoir site, it also led to changing fortunes downstream.

China's experience in South China forecasts future problems in Assam and Bangladesh if the SNWDP is successfully implemented. While India and China had agreed to share information on natural disasters as a part of the 1996 CBMs agreement (Article 8, Clause 2), the concerns are persisting after flash floods in Himachal Pradesh and Assam. India and China had signed MoUs on hydrological data exchange on the upper reaches of the Brahmaputra, although India now needs to pay hefty amounts for procuring such data. Although some Chinese water analysts at track II conferences the author attended in Beijing and Shanghai indicated that China is amenable to resolve any conflicts arising out of the use of these finite resources, specifically if these are directed towards irrigation purposes, this idea has so far not become an official stance and hence raises concerns in the lower riparian countries.

Fourthly, more specifically related to the Indus River and its tributaries, China's responses have been varied and contradictory in nature. On the one hand, China has stated that Kashmir is a disputed territory, but its developmental activities in Pakistan-controlled areas have clearly been tailored to the "all-weather" relations between China and Pakistan, not to the troubled relationship between China and India. Indeed, China has been gambling rather recklessly in this regard, since a majority of the investments of the promised $46 billion appear earmarked for the Pakistan-controlled Kashmir region. China also intends to secure the promised water storage, hydroelectric and infrastructure projects through the use of its own military forces in the medium to long term. Sending of Chinese military helicopters to Pakistan during natural disasters in Pakistan-controlled Kashmir region and the dispatch of Chinese security personnel to protect Chinese engineers and workers at construction sites are indicators of this possibility. The Indus basin unquestionably offers enormous scope for cooperation between China and Pakistan, partly due to the coincidence of strategic interests between the two and their joint opposition to India. On the other hand, China's responses to Indian requests for concluding an understanding on the Indus River have been met with much scepticism or opposition. Progress on this front has either been stalled or piecemeal. Thus we do not yet find basin-wide cooperation inclusive of China, India and Pakistan. While China takes notice of the Indus Waters Treaty between India and Pakistan, no such treaty exists between China and India, this in spite of the fact that the Indus originates in Tibet.

While cooperative efforts in relation to the Indus basin exist at a preliminary level in the form of MoUs, potential competition or even conflict clearly exists between China and India, mainly due to the former's balance-of-power approach towards New Delhi and India's sensitivities in regard to its sovereignty claims over Kashmir. India has objected officially to China's construction activity in the disputed Kashmir region. For an effective and integrated Indus River basin management, it is imperative that all the stakeholders in the basin area come together and explore options comprehensively. The experiment of the World Bank, which in 1960 brought together the rival states of India and Pakistan, is a useful example in this regard. Also, taking into account China's current focus on constructing hydroelectric dams in Pakistan, there is scope for exploring opportunities for transregional use of the Indus River system for transportation purposes by China, India and Pakistan. Innovative measures of this kind could lead to basin-wide cooperation as well as to economic interdependencies in the longer term that provide for an alternative and more promising paradigm.

References

"Hydroelectricity power plants in Pakistan". Accessed at http://www.industcards.com/hydro-pakistan.htm

"Rs101b scheme: Gilani inaugurates Mangla Dam expansion project". October 14, 2011 Accessed at http://www.awaztoday.tv/News_Rs101b-scheme-Gilani-inaugurates-Mangla-Dam-expansion-project_1_13698_Political-News.aspx

"Technical visit to Chashma Barrage and Jinnah hydropower project 18–19 February, 2012". Accessed at http://pecongress.org.pk/images/upload/books/6-30Technical%20visit%20Chashma%20Barrage%20and%20Junnah%20Hydropower%20project%20Final.pdf

"The Indus Water Treaty 1960". Accessed at http://wrmin.nic.in/writereaddata/InternationalCooperation/IndusWatersTreaty196054268637.pdf

"Water Law of the People's Republic of China'. 2002. Accessed at http://www.china.org.cn/english/government/207454.htm

AFP, China finishes canal to ensure water for olympics. February 26, 2008. *NewsEdge* Document Number: 200802261477.1_e6860033e4fd8291 File Number 985. Accession Number 258500521

Arpi, Claude (2004) The mysterious Tibetan Lake *Rediff.com* August 11, 2004 at http://www.rediff.com/news/2004/aug/11claude.htm

Bai, Minghua (2012) "印度河水争端解决机制的启示 – 兼论我国大湄公河水争端的避免解决" [Revelations of the Indus Water dispute settlement mechanism – Can China avoid and settle the Greater Mekong water dispute?] 南亚研究季刊 [South Asia Quarterly] No. 3 Issue 1550, March 2012, pp 101–106

Castelein, Saskia (ed) (2002) From conflict to co-operation in International Water Resources Management: Challenges and Opportunities, International Conference, 20–22 November 2002, UNESCO-IHE Institute for Water Education, Delft, The Netherlands

Central Tibetan Administration, Department of Information and International Relations, Environment and Development Desk (2000) Tibet: environment and development issues- water resources. Accessed at http://www.tew.org/tibet2000/t2.ch2.water.html

Chellaney B (2011) Water- Asia's new battleground. Harper Collins, New Delhi

Chen H, Rieu-Clarked A, Wouters P (2013) Exploring China's transboundary water treaty practice through the prism of the UN watercourses convention. Water Int 38(2):217–230, http://dx.doi.org/10.1080/02508060.2013.782134

Chengdu Business Report (2010) 印度河成第二个克什米尔" [Indus River becomes another Kashmir?] 成都商报 March 31, 2010. Accessed at http://www.apabi.com/nlc/?pid=newspaper.article&metaid=nw.D510100cdsb_20100331_2-14&wd=&cult=CN

China International Water Law. List of China's major international watercourses by country with treaties. Accessed at http://chinainternationalwaterlaw.org/pdf/treaties/PKW_list.pdf

China's Ministry of Water Resources at http://www.mwr.gov.cn/english1/20040801/38041.asp

China's Ministry of Water Resources, 800 million cubic meters south-north water delivered to Beijing. December 7, 2015 at http://www.mwr.gov.cn/english/Medianews/201512/t20151207_727309.html

Economy EC (2004) The river runs black. Cornell University Press, Ithaca

Economy, Elizabeth C (2008) Asia's water security crisis: China, India, and the United States. In: Strategic Asia 2008–2009: challenges and choices, September 2008. Accessed at http://www.nbr.org

Elhance AP (1999) Hydropolitics in the third world: conflict and cooperation in international river basins. United States Institute of Peace Press, Washington, DC

Geng, Shengwei (2012) 中印跨界水资源问题日益升温及其对中国的影响 [Growing Sino-Indian trans-boundary water resources problem and its influence on China]. MA dissertation submitted to Hebei Normal University, April 2012

Gleick PH (1993) Water and conflict: fresh water resources and international security. Int Secur 18(1):79–112

Gulati, Monish (2014) Water security at SCO; the omnipresent Chinese factor. October 4, 2014. accessed at http://southasiamonitor.org/detail.php?type=sl&nid=9149

Haider Kamran, Natalie Obiko Pearson (2015) Even China won't finance this dam as water fight looms. April 23, 2015. Accessed at http://www.bloomberg.com/news/articles/2015-04-22/even-china-won-t-finance-this-pakistan-dam-as-water-fight-looms

Hodum Ryan (2007) Conflict over the Brahmaputra River between China and India. ICE case studies, Number 205, May 2007. Accessed at http://www1.american.edu/ted/ice/brahmaputra.htm

Hong, Yang, Luan Shengqi (1998) "中国二十一世纪的水安全问题" [Problems of China's Water Security in 21st century] 中国环境管理 [China Environment Management] (1998) No. 4, pp 4–7

Hu, Wenjun, Yang Jianji, Huang Heqing (2010) 印度河流域水资源开发利用国际合作与纠纷处理的经验及启示" [Lessons and Inspirations from international cooperation and dispute resolution on water resources development in the Indus River Basin] 资源科学 [Science of Resources] No. 32, Issue 10 October 2010, pp 1918–1925

Huang, Zhiqiang (2010) 亚历山大对印度河流域的征服及影响" [The impact of the conquest of Alexander on the Indus Valley]. J Mudanjiang Normal Univ Issue 160, No. 6, pp 64–66

Huang, Lei (1999) 从巴基斯坦排灌渠工程- 看项目前期工作的重要性" [Irrigation channel projects in Pakistan – the importance of project preparatory work] 国际经济合作 No. 6, pp 51–53

Indian Central Water Commission See "Water and Related Statistics" April 2015. Accessed at http://www.cwc.gov.in/main/downloads/Water%20&%20Related%20Statistics%202015.pdf

Institute for Defence Studies and Analysis (2010) Water issues In India-China relations Chap. 3 in water security for India: the external dynamics. IDSA, New Delhi. At http://www.idsa.in/sites/default/files/book_WaterSecurity.pdf

Li Z (2015) 水与中国周边关系 [Water and China's Neighbourhood Relations]. Shishi Publications, Beijing

Li, Zhifei (2011) 中国与周边国家跨国界河流问题之分析" [An analysis on the issues of trans-boundary rivers between China and the periphery countries]. 太平洋学报 [Pacific J], pp 2–22

Li, Yunhui, Chen Xianyun, Sha Haiming (2011) "印度河流域水资源一体化管理框架研究" [Integrated framework for river basin water resources management in India]. 水利水电快报 32(9):5–8

Liu, Changming (1998) Environmental issues and the south-north water transfer scheme. The China Quarterly (Special Issue: China's Environment), 1998, no. 156: 899–910

Magsig B-O (2015) Water security in Himalayan Asia: first stirrings of regional cooperation? Water Int 40(2):342–353

McGaffrey, Stephen G, Mpanzi Sinjela (1998) The 1997 United Nations convention on international watercourses. Am J Int Law 92, no. 1 (January 1998): 97–107 as cited in Menniken, 2007

McGormack G (2001) Water margins: competing paradigms in China. Crit Asian Stud 33(1):5–30

Menniken T (2007) China's performance in international resource politics: lessons from the Mekong. Contemp Southeast Asia 29(1):97–120

Ministry of Water Resources, Government of India, India-China Cooperation. Accessed at http://wrmin.nic.in/forms/list.aspx?Id=4&lid=349

Ministry of Water Resources, Government of India, India-China Cooperation. Annual Report 2013–2014. Accessed at http://wrmin.nic.in/writereaddata/AR_2013-14.pdf

Mountjoy Shane (2005) The Indus River. Chelsea House Publishers

Mustafa Khalid Tarbela Dam to produce 6,200MW electricity by 2018. The News March 8, 2015. Accessed at http://www.thenews.com.pk/Todays-News-6-305635-Tarbela-Dam-to-produce-6200MW-electricity-by-2018

National Development and Reform Commission (2006) Hydropower in the west: priority for China,November30,2006.NewsEdgeDocumentNumber:200611301477.1_59500031c568ae45 File Number 985 Accession Number 235850374

Nickum James E, The status of the South to North water transfer plans in China. Accessed at http://hdr.undp.org/sites/default/files/james_nickum_china_water_transfer.pdf

Osborne Milton (2015) Will China's dams control the Mekong's flow?. February 16, 2015. Accessed at http://www.lowyinterpreter.org/post/2015/02/16/Will-Chinas-dams-control-the--Mekongs-flow.aspx

Pakistan Water and Power Development. Authority at http://www.wapda.gov.pk/htmls/power-index.html

People's Daily (2007) "世界十大河流面临严重危险" [World's top ten rivers in danger], November 15, 2007

Reuters (2006) China denies plan to dam Brahmaputra river, November 22, 2006. Accessed at http://timesofindia.indiatimes.com/articleshow/522947.cms

Sina.com (2013) "2013年4月中印对峙地点仍在中国境内" [The April 2013 standoff area between China and India is still within China], April 28, 2013. Accessed at http://blog.sina.com.cn/s/blog_48df91ed0101co06.html

Sud, Hari (2008) "China's future water war with India" May 13, 2008. Accessed at http://www.upiasia.com/Security/2008/05/13/chinas_future_water_war_with_india/3300/

Tate E et al (2001) "受泥沙威胁水库的模拟管理" [Simulation management of threats to the reservoir from sedimentation] 水利水电快报 [Water Resources & Hydro-power Information] No. 22 Issue 18 September 2001, pp 12–15

The Dawn (2015) "Another $51m to expedite Tarbela project completion" March 7, 2015. At http://www.dawn.com/news/1167908

The Hindu (2004) "Flash-flood threat hits tourism" August 18, 2004. At http://www.thehindu.com/2004/08/18/stories/2004081806240300.htm

The Nation (2010) "172 Chinese engineers shifted from Chashma", August 2, 2010. Accessed at http://188.138.70.147/politics/02-Aug-2010/172-Chinese-engineers-shifted-from-Chashma

The Nation (2009) "Chinese firm to build Bunji Dam" August 23, 2009. Accessed at http://nation.com.pk/politics/23-Aug-2009/Chinese-firm-to-build-Bunji-Dam

The Nation (2013) "Wapda inks MoU for 1410MW Tarbela 4th Extension project" September 10, 2013. Accessed at http://nation.com.pk/business/10-Sep-2013/wapda-inks-mou-for-1410mw-tarbela-4th-extension-project

United Nations Food and Agriculture Organisation, "Indus Basin". Accessed at http://www.fao.org/nr/water/aquastat/basins/indus/index.stm

Vinayak, Ramesh (2005) "Sutlej in Himachal swells as Pareechu lake bursts in Tibet". India Today, July 11, 2005 at http://indiatoday.intoday.in/story/sutlej-river-in-himachal-pradesh-swelled-up-after-pareechu-lake-burst-in-tibet/1/193302.html

Wang, Xiaoyang (2013) "中印跨界水资源问题研究" [A study on the Sino-Indian trans-boundary water resources] MA dissertation submitted to Central China Normal University, May 2013

Wolf AT, Natharus JA, Danialson JJ, Ward BS, Pender JK (1999) International basins of the world. Int J Water Resour Dev 15(4):387–427, p.401

Xinhua (2006) "Yellow river's thirst for water brings China's shortages into light March 23, 2006 *NewsEdge* Document Number: 200603231477.1_619000520bcdd212 File Number 985 Accession Number 223250038

Xinhua (2007) China tunnels through yellow river for massive water diversion Project December 28, 2007, NewsEdge Document Number: 200712281477.1_0b4b004160fa76a2 File Number 985 Accession Number 255500454

Xinhua (2007) "Restoration at Yellow River Source Aims To Curb Desertification" December 27, 2007. Accessed from United States Department of Commerce, Foreign Broadcast Information Service, NewsEdge Document Number: 200712271477.1_0bb5003698f730d7 File Number 985 Accession Number 255450191

Xinhua (2008) China to speed up building gigantic South-To-North water diversion project in 2009 December 15, 2008 NewsEdge Document Number: 200812151477.1_151700434103f4ed File Number 985 Accession Number 273100659

Yue Z (2004) Zhongguo shuili fazhan zhanlue wenji (1996–2004) [Collection of China's Water Development Strategies (1996–2004)]. Zhongguo shuili shuidian chubanshe, Beijing

Zaffos, Joshua (2014) "Life on Mekong Faces Threats As Major Dams Begin to Rise" February 20, 2014. Accessed at http://e360.yale.edu/feature/life_on_mekong_faces_threats_as_major_dams_begin_to_rise/2741/

Zhong, Huaping, Li Jianqiang, Wang Jiansheng (2011) "印度河与印巴用水问题研究" [A Study on the Indus River and the problems of water sharing between India-Pakistan] 世界农业 [World Agriculture] No. 382, pp 68–70, February 2011

Part IV
Concluding Perspectives

Chapter 10
The Role of International Development in Reimagining the Indus Basin

Zafar Adeel and Paula R. Newberg

Abstract The states in the Indus basin region have benefited from international development assistance to varying degrees; however, the history of these initiatives points to approaches that do not consider the basin as a whole or a single unit. This disparity between the social, economic, and technological needs at the basin scale and responses of the international community have as much to do with internal conflicts and discord as with the lack of a cohesive vision on the part of the external interlocutors. This chapter unpacks some commonly made assumptions about engagement of international partners and analyzes how these assumptions broadly fall apart for the Indus basin due to a multitude of external and internal drivers. New development paradigms, emerging from the 2030 Agenda for Sustainable Development and the 2015 Paris Agreement on Climate Change, are assessed for their potential to reshape regional development patterns. It is concluded that regional approaches – which fully engage Afghanistan, China, India, and Pakistan – are the best mode for moving forward to cope with existing and emerging challenges around economic growth, water, food, and energy security; and health and well-being of people living in the Indus basin. The international community can proactively foster such new relationships, but only if the four states are amenable to accepting such interventions.

Keywords International development assistance • Foreign aid • International community • Water governance • Water-food-energy security • Regional coordination

---

Z. Adeel (✉)
United Nations University, Institute for Water, Environment and Health, Hamilton, ON, Canada
e-mail: adeel_unu@yahoo.com

P.R. Newberg
University of Texas, Austin, USA
e-mail: pnewberg@austin.utexas.edu

Z. Adeel, R.G. Wirsing (eds.), *Imagining Industan*, Water Security in a New World, DOI 10.1007/978-3-319-32845-4_10

10.1 Background and Contexts

The Indus basin is by its nature a region, a transborder interest group, and an artifact and critical element of the international political economy. While international engagement has been an assumed element of its identity and trajectory, this has been limited to the two primary states for more than 50 years – India and Pakistan. Meanwhile, each state in the basin continues to encounter serious tests of its traditional security, military engagements, development choices, and economic and diplomatic relationships – with one another and outside the region. No state in the region – or in an imagined Industan – lives apart from several (or all) of the others, and all are engaged in crucial bilateral and global institutional relationships that, in many ways, are the foundation for the Indus catchment's politics, societies, and economies. Equally important, the relationships among the states that comprise the basin and the paths that the Indus traverses hold critical places in contemporary international politics and in the evolving response systems for climate change.

For this reason, international engagement is not merely a choice for the future of the Indus basin. Rather, it is the essential policy context within which the region will develop. The most critical issues of engagement, therefore, are about how the region and its members comport themselves, individually and collectively, in the international arena – with regard to water security in the first instance and the constellation of issues and challenges that face them all as well. In this sense, the Indus basin is both an experiment and a challenge for the idea of international engagement in the early twenty-first century: it is at once embedded in political histories that are difficult to transcend while at the same time living at the cusp of decisions about the future of the region and its waterways that will affect the health, welfare, livelihoods, and existence of these states and their people.

International engagement, for the purposes of this chapter, is therefore treated as a two-pronged idea. In the first instance, it revolves around habitual and deeply held understandings about the political capacities, interests, threats, and prospects for each state and its neighbors. Second, and more important for this discussion, it is also premised on the significant role that international assistance plays in providing space for experiments and investments in regional cooperation through material and financial resources and political impetus, provocation, and protection.

The states in the Indus basin catchment are developing and transitional. All encounter a general lack of capacity at multiple levels of the water sector – human, institutional, technological, and service provisioning – that underpins their inability to adequately respond to water and other related challenges (UNDP 2006; Mustafa 2007). The lack of resources refers primarily to financial resources and their allocation, with the assumption that the challenge in mobilization of finances is not absolute but is relative when compared against other competing priorities of the respective governments. These resource and capacity gaps lead to a level of ineffectiveness in understanding the water challenges, analyzing the drivers and causes, identifying adequate solutions, and achieving the successful implementation of policies and practices that can benefit the entire region (Mustafa 2007).

Numerous publications have analyzed how the lack of capacity and resources in the Indus basin correlates to inadequate governance mechanisms and leads to a persistent downward spiral of poverty and poor well-being (Ali 2002; Mustafa 2007; NIC 2012). The analysis in this chapter focuses on the role actors from outside the region have played in the water domain and how they might influence outcomes for overcoming water-related challenges. Collectively, these external actors – comprising development banks, bilateral donor agencies, United Nations (UN) organization and agencies, internationally driven nongovernmental organizations, and academic and research organizations – are characterized as the "international community." To state the obvious, this notion of international community is amorphous and is often used differently in different contexts. For the sake of our analysis in this chapter, we will use the definition provided here.

10.1.1 Underlying Assumptions for International Engagement

This section focuses on the institutional and governance contexts in which the international actors engage with developing countries. It explores some common assumptions made about this international engagement and unpacks their validity in situations commonly encountered. In particular, it looks at the current agenda for engagement in order to clarify current and emerging governance issues that figure in future water-based policies and decisions. These issues, while common to the entire scope of international engagement, are particularly salient in discussions about foreign assistance and investment. This generic analysis is meant as a device to reinforce subsequent discussions about what international engagement might mean in the future for resolving current and emerging challenges in the Indus basin and how one drives the other. It does not attempt to replicate detailed analysis undertaken elsewhere of the linkages between international assistance and governance, either as a criterion for receiving aid or as an outcome of aid-based development (e.g., Mosley et al. 2004; Copestake and Williams 2014; Michaelowa and Weber 2007).

The effectiveness of international engagement, driven through assistance, loans, and, in some cases, foreign direct investment (FDI), can be measured quantitatively over time and linked to a qualitative development narrative. Conventional wisdom also demands that aid effectiveness can only be delivered by recipient "partners" that are "committed to common goals and capable of absorbing extra resources" (Copestake and Williams 2014). We describe here four key assumptions in this argument and identify reasons why these assumptions may not hold out, either partially or completely, in many cases.

First, it is assumed that international assistance follows the five core principles laid out in the 2005 Paris Declaration on Aid Effectiveness: ownership, alignment, harmonization, managing for results, and mutual accountability (OECD 2006). While the Paris Declaration clearly lays out generic principles for development assistance and how it can be better harmonized and integrated, it also recognizes a number of challenges that have to be overcome. The latter include failure to provide predictable multiyear commitments, insufficient integration of development initiatives into recipient countries' broader development agendas, and overcoming corruption and lack of transparency (OECD 2006). One may argue that the Paris Declaration has resulted in some improvements but the challenges are far from being overcome in the 10 years since. Chan and Chung (2015) argue that failure to achieve many Millennium Development Goals (MDGs) by 2015 may be indicative of more fundamental underlying failures of development assistance. The responsibility for this inability to meet commonsense principles could be apportioned in equal measure to both donors that often let political considerations trump the Paris principles and recipient governments that are not able to provide a policy environment conducive to success.

Second, it is often assumed that international assistance is neutral in its nature, but this assumption relies on context and level of analysis. This assumption is most difficult to analyze in a quantitative and systematic way because the "strings" attached to many forms of international development assistance are not publicly disclosed. Most major donors have strategic interests in the region that go beyond the specific programs and projects they fund and tactical concerns that often drive decisions about specific interventions. For example, the testing of nuclear devices by India and Pakistan in 1998 led to significant but short-lived sanctions on Pakistan and India, including those imposed by the United States and a number of OECD[1] countries (Morrow and Carriere 1999). Similarly, many bilateral aid programs are tied to technical and technological assistance coming exclusively from the donor countries, lowering their effectiveness (Mosley et al. 2004); this approach of "tied aid" is very rarely publicly acknowledged, as was done by Canada in 2012. Some even argue that a large fraction of the assistance ends up back in the donor countries; an Afghan expression used in this context is "cows that drink their own milk" (Suhrke 2006). Yet another trend is to bypass governmental channels and utilize local and international nongovernmental organization for the delivery of aid; this was prominently observed in the aftermath of the Asian Tsunami of 2004, when donor governments also demonstrated reluctance in fully disclosing their activities to recipient countries (Kelegama 2012). Collectively, international aid is often driven by a combination of economic, social, commercial, and geostrategic agendas.

Third, it is assumed that recipient countries can demonstrate quantifiable changes in their governance mechanisms and related institutions in a well-defined time frame. However, a common underlying challenge is that aid and development loans are often perceived as transfers or grants with repayment beyond the tenure of individual governments, resulting in less emphasis being placed on evaluating the long-term need for them or setting up institutional mechanisms for achieving effectiveness (Kelegama 2012). A considerable component of the success of international assistance is the effectiveness of governance and related institutions in place in the recipient country (Michaelowa and Weber 2007; Burnside and Dollar 2000).

[1] OECD: Organization for Economic Cooperation and Development.

Notably, Pronk has argued that instead of using better governance as a precondition for receiving development assistance, it should be the primary objective (Pronk 2001). Interestingly, an analysis of Dutch development assistance to aid recipients, including India and Pakistan, suggests that Dutch policymakers were unable themselves to meet the governance criteria they wished to apply to recipient countries (Hout 2002). In a report published by the Danish Institute for International Studies, Ravnborg et al. (2013) argue that there has been a trend of increasing investments and activities focusing on environmental governance. Whether this trend truly reflects the governance-first approach suggested by Pronk is an open question.

Fourth, there is a common assumption that international engagement can bring otherwise recalcitrant stakeholders to the table and achieve mutually agreeable solutions to basin-wide problems. The adoption of the Indus Waters Treaty is often cited as an example in which the World Bank stepped in as a mediator between India and Pakistan (Mustafa 2007). In general, it has been argued that the United Nations system brings a measure of unbiased and apolitical implementation of development assistance, particularly because it is not financing big infrastructure projects and in some cases is not directly promoting economic growth (Neumayer 2003). Ergo, the UN system ostensibly can be more neutral in providing policy advice to government, but in reality it is not entirely immune to being torqued by the donor countries. In more extreme situations, donor agencies tend to bypass established UN organizations and agencies altogether; an example of this is the case of worldwide development assistance for health in which the international community has shifted away from the UN and development banks to direct targeted aid through more specialized institutions like GAVI[2] (Ravishankar et al. 2009). One may conclude from the foregoing that the Indus basin requires a new kind of multilateral engagement that is regional in a sense and issue specific in another sense.

10.1.2 International Engagement Experience in the Indus Basin

Let us analyze the patterns of international engagement in the Indus basin, particularly in Afghanistan, India, and Pakistan. The engagement of China in the Indus basin differs on a number of levels – the relatively miniscule fraction of the watershed falling in China, its rapid economic growth which has changed its status as aid recipient, and historical lack of significant claims on the headwaters of the Indus basin. The relationship between China and the other riparian countries – particularly noting a recent trend to construct hydropower projects in the upper reaches of the Indus basin – is discussed in depth by Kondapalli (Chap. 9, this volume); hence we have not included China as a riparian state in our analysis in this chapter.

[2]GAVI: The Global Alliance for Vaccines and Immunizations.

The donor community views Afghanistan, Pakistan, and India quite differently as potential aid recipients. Afghanistan has been in a state of chronic conflict since 1979, including two full-blown invasions by external military forces, interspersed with various stages of civil war (Chandra 2015). In these terms, it doesn't even fit the commonly used "post-conflict reconstruction" description because armed conflict has never been fully eliminated in the last three decades. Consequently, governance mechanisms are either nonexistent or very poorly developed, which poses a serious challenge to any international assistance regime. Further, lack of absorptive capacity in its institutions, particularly for rural settings, considerable level of corruption, and less-than-desirable operating conditions have been tied to a high level of ineffectiveness in aid delivery (De Silva 2012). As its gross national income (GNI) per capita stays below US$995, it remains in the low-income country category and is eligible to seek assistance for debt relief (Kelegama 2012). Its aid composition has also evolved since 2001: support to the security-related components of the government has declined, and there was a corresponding increase in assistance to education, infrastructure, and rule of law (De Silva 2012). Byrd (2016) further argues that the drawdown of international troops over the past few years has led to a "sharp economic slowdown" during the 2013–2015 period.

Pakistan has demonstrated considerable economic growth in the past three decades, even if this growth has been very uneven in a geographical sense and has occurred in irregular spurts with intervening periods of political instability. Based on its GNI, Pakistan receives a mix of concessional and non-concessional assistance from the development banks, notably the World Bank (Kelegama 2012). Governance mechanisms and democratic institutions are also uneven in their level of development and effectiveness, circumstances further compounded by capacity gaps in human, technological, and institutional domains (Briscoe et al. 2005). Concerns of the international community around security and the resulting volatility in aid provisioning have led to fragmentation and poor aid effectiveness (Ahmed and Wahab 2012). Conversely, there is a great deal of geopolitical interest in Pakistan, not the least because it is a nuclear-armed power but also because of its location at the crossroads of energy and trade corridors.

India claims to be the largest democracy in the world, with matching strength of governance mechanisms. It nonetheless is adversely impacted by strong disparities in the level of economic development. Like Pakistan it is also eligible for a mix of concessional and non-concessional assistance – although it is anticipated that grants and concessional loans will give way to loans with tougher conditions (Kelegama 2012).

10.2 Reimagining the Region for International Institutional Engagement

10.2.1 A New Agenda for International Engagement

There has been a major shift in the international development paradigm in late 2015 in the form of a new "2030 Agenda for Sustainable Development" adopted by world leaders (UN 2015). Many have characterized this as truly transformative because it is universal and applies to all countries; it is comprehensive and seeks to completely eliminate problems; it is complex as shown by the large numbers of goals and targets; and, it is ambitious because it aims to fix major global problems in a 15-year span. The range of development issues covered by the resulting sustainable development goals (SDGs) is quite broad, but one may argue that water security remains at the core of targeted economic, social, and environmental progress. In addition to a dedicated goal (SDG 6) on universal water issues, other SDGs – such as those related to health, cities, consumption, marine resources, and terrestrial ecosystems – also include water-related targets.

The other component of the paradigm shift is focused on climate change; the global leadership gathered in Paris in December 2015 and hammered out a new climate change agreement that sets flexible and revisable targets for the reduction of greenhouse gases and allocates resources for increased resilience and adaptability against climate-related impacts on economies and societies. Many have heralded the Paris Agreement as a major achievement and success, while naysayers argue that it does not go far enough in cutting greenhouse gas emissions. In either case, the success of the agreement's stipulations will be driven by how parties to the agreement reshape their own economic and social development plans. For the Indus basin states, climate change portends an increase to the water-related challenges faced by the governments; Michel (Chap. 6, this volume) describes how collective actions can help improve the level of understanding around these challenges and the capacity for making effective policy choices. Yet another element of the global dialogue around international development was the Financing for Development Summit that took place in Addis Ababa in July 2015. It yielded mixed results, which many found lacking in terms of not being able to identify visionary solutions for financing future sustainable development.

It is still early in the day to determine the full impact of this new 2030 Agenda on the international development paradigm, but a few assertions can be made safely. First, there will be a greater focus on removing inequalities; while this is meant to primarily focus on inequalities within countries, it is conceivable the inequalities at the regional or river basin scale in transboundary settings would also be of interest. Whether this emphasis on achieving greater equality in access to water resources creates much greater demands on often dwindling freshwater resources, or helps manage the water demand more effectively to match available resources, remains an open question. By the same token, the adverse trend in inequality – for example, the rapidly widening gap in global wealth distribution – can only be reversed with great difficulty (Cingano 2014).

Second, financial capital for meeting these commitments – which would include investments into infrastructure development, establishment of governance institutions, building human and technological capacity, and initiatives that spur economic and social development – will have to be found primarily within countries, considering that the scale of capital needed often exceeds international aid by at least one order of magnitude (Schuster-Wallace and Sandford 2015). The "domestic" financial capital could come from either public or private sources; the latter would likely require considerable shifts in enabling policies that protect private investments. Taken at face value, this implies a somewhat diminished, but still very important, role for international development assistance.

Third, while the importance of adequately managing water resources is underlined through well-defined targets in a number of SDGs, whether governments are able to develop cohesive policy responses remains to be seen. Based on a 2012 survey of more than 130 countries, many countries were struggling with the integration of water management across various economic sectors (UNEP 2012). Unless governments in Afghanistan, Pakistan, and India receive substantial international assistance in reformulating national development policies, it is unlikely that the implementation of SDGs will change their planning processes in the short run. It can be argued that this is an opportunity for the international community to step in and help expedite integration of national budget planning with sustainable development and climate change targets.

The success of this new international development agenda will be driven by the ability of governments to break down, or at least undertake a major transformation, of the sector-based silos in national planning. The presence of policy incentives for greater integration of approaches and creation of apex institutions within governments would have to be the first steps in that direction (Schuster-Wallace and Sandford 2015). Historically, international interventions – notably by IMF and development banks – for achieving such policy integration have met with only limited success.

10.2.2 Necessary Elements of Future Regional Cooperation

A number of existing regional-scale institutions can be invoked to reimagine an "Industan" in which governance of water resources can be linked closely to broad rights-based governance across borders. Such political realignment has to go hand in hand with broader agendas for regional integration. Ali and Zia (Chap. 7, this volume) argue that existing regional organizations offer a way of overcoming historical inertia; they specifically discuss the potential for the South Asian Association for Regional Cooperation (SAARC), which has included Afghanistan since 2007, to play a bridging role in bringing Indus basin riparian states to the table. A specific suggestion offered by Ali and Zia is to increase data sharing, which will help reduce data gaps and uncertainties, while also offering an alternative channel for trust building. A number of other international organizations can further play a role in regional cooperation. Most notably, the Shanghai Cooperation Organization (SCO) offers a platform for engaging China, particularly because India and Pakistan will soon join the coalition as full members (Afghanistan retains an observer status). The SCO can also broaden the scope of cooperation beyond shared water governance to include partnerships around energy sharing and transport cooperation.

Some regional organizations that are focused on more technical and scientific cooperation are yet another mode for regional cooperation. Supported by the World Bank Group and a number of other donors, the International Centre for Integrated Mountain Development (ICIMOD), the International Center for Agricultural Research in the Dry Area (ICARDA), and the International Crops Research Institute for the Semi-Arid Tropics (ICRISAT) are all quite active in the Indus region and can be invoked to undertake region-scale cooperation around water and food security.

10.3 New Institutional Assumptions and Mechanisms

10.3.1 Emerging Trends in International Assistance

As noted earlier, the five principles associated with the 2005 Paris Declaration (together with the subsequent 2008 Accra Agenda for Action) have emerged as the gold standard for international development assistance. A number of researchers have reviewed their impact on aid delivery on the ground, assessing whether measurable improvements have actually taken place (Klees 2010; Chan and Chung 2015). Kaufmann (2009) argues that, at a minimum, provisioning of aid to extremely poorly governed countries has been curtailed, and as a result, "some actual progress in untying aid, conditionality, and capacity development is already apparent in practice in some places." However, he contends further that while there is a push for increased transparency, there is little effort made to improve governance or to curb corruption. It does not help that the 2008–2009 global financial crisis has also exposed governance and corruption problems in developed countries, undermining the moral authority for making conventional demands for "good governance" of aid-recipient countries.

There is yet another, larger metamorphosis taking place as newer "donors" like China, Malaysia, and India have entered the international development space (Tilak 2014; Chan and Chung 2015). However, these emerging donor countries are not beholden to the principles followed by OECD DAC[3] – as they are not members of that group – and in many cases have no clear description of an assistance policy. A counteracting "Beijing Consensus" puts much greater focus on strong, government-based initiatives than the more traditional linkages to free market economy (Yao

[3]OECD DAC: The Organization for Economic Cooperation and Development's (OECD) Development Assistance Committee (DAC).

2011). This is further coupled with the notion of strengthening geopolitical alliances as an aid objective. One may argue that these new multifaceted and multidimensional approaches to international assistance adopted by the emerging donors could lead to a range of possible outcomes: for example, increasing dysfunction in international assistance as OECD and emerging donors operate at cross-purposes, or eventual mutual alignment and re-shifting of aid policies by both emerging and OECD donors, or a more lopsided influence on the approaches being used by OECD donor countries, or some combination of all of the above.

It can be argued that the shifting dynamics of international development assistance offer the opportunity to the Indus basin states to demand that such assistance must cater to their own social and economic imperatives (such as economy of scale in development assistance) and also take into account their proximity to fast-growing economies in the ASEAN[4] region (Cho et al. 2014). This would require collective visioning and planning at the basin scale; the regional instruments' discussion in Sect. 2.2 can be invoked to facilitate a dialogue. At the same time, there is mounting evidence that development assistance – while creating long-term "moderate" benefits – is not a panacea for all development challenges (Arndt et al. 2011).

10.3.2 Regional Realities

During the past decade, international assistance to Afghanistan and Pakistan has assumed an additional dimension of being tied to the "war against terrorism" and achievement of regional security. In fact, a vast majority of the assistance from United States to Pakistan has been tied to "short-term counterterrorism objectives," with only about ten percent of the aid going toward development and humanitarian assistance. The situation is considerably worse in Afghanistan, with a prevailing view that neither international assistance nor the security apparatus have achieved stability or critical mass to result in long-lasting outcomes. This situation leads to obvious impacts on all international assistance aimed at these two countries. On the other hand, India offers considerably greater political stability, although that has not automatically translated into reduction in national poverty levels.

Governance challenges at national and subnational levels are ostensibly the greatest hindrance in achieving sustainable economic development in the Indus basin states. These shortcomings not only make it challenging to interface with partners from outside the region, but it becomes nearly impossible to achieve aid effectiveness. An additional dimension is the absence of vertical integration of governance and planning at the national level, which leads to fragmentation of international assistance and in extreme cases donors walking away. This phenomenon is more pronounced in Afghanistan and Pakistan, where provincial autonomy and national policy integration often work at cross-purposes.

[4]ASEAN: The Association of Southeast Asian Nations.

10.3.3 Mobilizing Finances at the Indus Basin Level

Mobilization of development assistance comprises two broad elements: international assistance in the form of loans or grants and resources allocated as part of routine budgetary and fiscal planning at either national or provincial levels, or both. We have discussed that the former is limited in its scope and increasingly competing with other "geopolitical security concerns." Additionally, the scale of resources available from without are often inadequate for addressing challenges at the national scale; some have argued that these are merely "seed funds" aimed to guide national economic and development policies (Schuster-Wallace and Sandford 2015). The latter element requires considerable attention in retooling national policies to better address the combined food, energy, and water challenges in the Indus basin.

In addition to more conventional resource mobilization, a number of innovations have also emerged in the last decade which supplement, and often integrate with, the two traditional modes of resource mobilization discussed above. These can be categorized as four broad approaches (Schuster-Wallace and Sandford 2015): First, removal of adverse subsidies, including those provided to the energy sector that are counterproductive, could release significant resources for economic growth and infrastructure development. The global scale of these subsidies is about US$ 1.9 trillion per year (World Bank 2013). An estimate for the Indus basin region is not readily available; as an example the electricity sector subsidies in Pakistan are estimated at over 2 % of the national GDP (Walker et al. 2014), and the fertilizer subsidies are estimated at about Rs. 27 billion (ca. US$ 250 million) per year (Valdés 2013). Second, reduction of corruption can free up as much as a third of the investments typically made in the water sector (Transparency International 2008). Third, the private sector and local entrepreneurs can be a resource for capital, given the appropriate and enabling policy environment. This has become particularly important in the face of declining international aid after the global financial crisis of 2008/2009. Many long-term investors like pension funds, insurance companies, mutual funds, and sovereign wealth funds can be a part of this equation (Schuster-Wallace and Sandford 2015). Fourth, an evolving and growing resource for capital is private philanthropy; numerous models that utilize crowdsourcing or crowdfunding approaches have cropped up in recent years (Agrawal et al. 2013).

We argue that given broader considerations around governance challenges in the Indus basin, some of these more innovative capital mobilization approaches coupled with government allocations and/or international assistance may stand a good chance of success. Pushing this argument even further, it may be argued that enabling local entrepreneurs ought to be a key objective of international assistance.

10.4 International Institutions as Trusted Interlocutors

10.4.1 Evolving Regional Development Paradigms

The context of politics and, by extension, economics in the Indus basin region has long been conflict rather than cooperation, and competition rather than collaboration. Cross-border tensions and often, outright cross-border wars, as well as intrastate tensions, have had three effects on the way that outside actors have looked at the region's political economy: first, outside actors have been more inclined to look at the region as a medium for gaining their own tactical and strategic advantages rather than in consideration of the needs of local actors; second, and in part as a result of persistent tensions, development relationships have been almost entirely intrastate in means and ends; and third, political relationships have been managed incrementally in small, manageable steps rather than guided toward a larger vision. The last has been true of discrete interventions – for example, easing tensions between India and Pakistan over a variety of issues in the last decade – as well as of the relatively successful legal and institutional management of the Indus Waters Treaty. It has also meant that dispute resolution has taken primacy over conflict prevention – whether about conflicts over the Indus or more general conflicts between and among Indus basin states – and national goals over shared goals. In effect, the Indus basin has modeled "the tragedy of the commons" in a space and policy arena in which shared interests should dominate.

What should this mean for future hydro-diplomacy, and how can external actors help local actors to change direction? Are those external actors who have been most relevant in recent decades likely to be most effective in championing the interests of Industan, and are new regional players likely to take on different roles? The analyses in this volume and in this chapter suggest several ways to think about these questions from the point of view of an imagined – and, thus, regional – Industan.

First, broadening the roles of Industan's stakeholders means broadening the number and kinds of stakeholders in the region. From the point of view of policy, this is both inward and outward looking: tensions over water access and management within states and among them will have to be reconciled, as neighboring provinces in Afghanistan and Pakistan, and Pakistan and India already illustrate. Additionally – and critically – relationships among states in the region need to be calibrated to further common goals: mutual rather than differential investments and financing and intra-regional assistance that forwards shared interests of water access and quality, management, and sustainability.

Second, Industan needs to see itself as a region, not a collocation of individual actors. As such, its interests are distributed, not divided. This can be a difficult leap to make, given decades of interventions premised on the accretion rather than the circulation of power. If, however, the international community views Industan's challenges as opportunities to broaden their own engagements in the region, they can support the region to reformulate its political and economic identities. In essence, distributed risk within the region, underscored by supportive actions from outside actors, can both establish and maintain new power relationships that maximize cooperation. This approach can equally apply to private sector investment, although perhaps much harder to enforce in practice.

Third, the notional identity of Industan, if translated into revised practices of aid as investment, can help to further a more stable region in political terms and to alter the old categories of political and economic interventions. Some have argued that the old model is the World Bank or the United Nations, and the new model is the AIIB or newer, potentially more flexible institutions like the SCO (Lim 2015). Distinctions notwithstanding, the more important issue is recalibrating financial relationships in aid of broader political relationships that emphasize procedural equality and political equity. This is a hard road to travel: the sheer differences in size, wealth, and power within the region can seem insurmountable – particularly when one actor, China, has embarked on enormous infrastructural investments in the region (bilaterally and through the AIIB) that appear to counter the equities embedded in an imagined Industan. At the same time, China's largely unilateral investment plans for regional roads – which clearly are in aid of Chinese rather than regional business interests – provide an analogy for Indus water management by seeking to enhance it through regional investments.

10.4.2 Brokering Relationships

To encourage and facilitate these relationships will require new and renewed commitments on part of the international community to forward-looking diplomacy. That, too, will be a challenge, since it is easy for the disputes embedded in the region to overtake new initiatives.

Again, there are two ways to think about this problem. The first is to reinvigorate multilateral institutions (on behalf of the region) by taking their explicit mandates – sustainability, development, peace, and security – as the basis for the region's collective human security. The second is to entrust bilateral actors, whether aid providers and aid recipients, or conflict parties and conflict resolvers, to further this agenda based on shared principles. Both assume a commitment to a general welfare premised on the health, vitality, and future of the Indus basin; both, in the end, assume collaboration among all actors, regardless of provenance.

Most importantly, however, the impetus for preventing conflicts over water and creating a stable basis for water sharing and investments in water infrastructure must come from the basin itself. Each state in the region is a regional and international political and economic actor. To the degree that enmity overrides amity, these states may accept the honest efforts of international brokers. History suggests that new models of cooperation are often crafted as a result of a major crisis; for Inustan, these might include the threat of droughts, floods, famines, and disease. The mechanics of such cooperation and collaboration, and supporting international aid and investment, can only be a small piece of a much broader political and international commitment to save the Indus for those whose lives depend on it.

10.5 Imagining the Future of International Cooperation in Industan

International assistance is one domain in which it is feasible and practical to visualize a scenario where basin-wide initiatives can be developed and fostered. In some ways, it harks back to the concepts introduced by David Lilienthal in 1951, with three key modifications. First, in addition to serving as an honest broker, the international community can also provide the human, financial, technological, and institutional resources for making change happen on the ground. Achieving such an overarching plan would require an integrated and all-inclusive exercise to arrive at a vision that serves multiple objectives. These objectives would include achievement of water, energy, and food security while building sufficient goodwill and trust among the riparian countries.

Second, this effort would need to be comprehensive in its geographical scope and thus include all four riparian countries. While this could ostensibly amplify conflicting political positions of the states involved, we argue that an ambitious approach is exactly what is needed to bring cohesion in an otherwise politically fragmented situation. Recently established energy-based coalitions are a case in point; for example, the 1000 Electricity Transmission and Trade Project for Central Asia and South Asia (aka. CASA 1000) between Kyrgyz Republic, Tajikistan, Afghanistan, and Pakistan, and the Turkmenistan-Afghanistan-Pakistan-India Pipeline (aka. TAPI). Given the wider interest in bringing political stability and economic growth to the region, an imaginative initiative could capture the attention of the international community as well.

Third, the approaches developed would have to be holistic and integrative. That means negotiating development projects around water alone will not suffice. Other key sectors such as food and energy would be essential for political and economic success of internationally assisted initiatives. It would also require engagement of all stakeholders, most notably the private enterprise that has been hitherto excluded from such dialogues.

A 360° review of the situation is also required to convince all parties to collaborate with the international community. The downside of not being able to arrive at a common vision and implementation frameworks must be weighed against the mutual benefits to be reaped through basin-wide initiatives.

10.5.1 Overcoming Roadblocks to Progress

The two major categories of roadblocks to imagining Industan and convincing the international community to fully support this vision are roadblocks internal to each of the riparian states and those that are external to the region.

As discussed in this chapter, the absence of adequate governance mechanisms and/or their effective implementation is by far the greatest obstacle that each country needs to overcome internally. Some of these governance challenges are systemic and go far beyond water (or energy, or food) management; collaborations around the Indus basin may serve as a trigger to improvements in governance and related institutions. Corruption is yet another major obstacle to achieving effective development – both by siphoning off resources that would otherwise serve the water sector and by making it more difficult to seek incremental resources for filling in institutional or infrastructure gaps. Subnational politics further complicate international engagement – and it manifests differently in each of the riparian states. The interprovincial rivalries are the most intense in Pakistan and have become a real obstacle in achieving national consensus around water and energy management. In Afghanistan, the situation is even worse when we consider the deteriorating security situation in most provinces and a likely return in coming years to the anarchic situation prevailing through most of the 1990s. This makes it challenging to develop plans that would apply across the board in the country. On the other hand, India seems to have a strong national consensus and plans for water management; the implementation of these plans typically falls short because of other governance challenges discussed earlier.

The Indus basin also has to cope with numerous externalities, most of which are not driven or controlled by the governments in question. In a positive sense, there is sufficient interest in creating regional energy corridors, which indirectly make it possible to discuss sharing of water resources; in some cases, sharing of water and energy are inextricably linked. The economic drivers for such collaboration are much wider than the footprint of the Indus basin. On a negative note, terrorism and other security concerns – including the presence of ISAF[5] and NATO[6] forces in Afghanistan, interest in the region's mineral and petro-carbon resources, and access to seaports in the Arabian Sea – are part of a larger landscape of geopolitical drivers that make it difficult to bring investments into the region. It could be considered a "catch-22 situation" in which external investments and international aid can bring about political stability and economic growth, but doing so requires a base level of these elements.

[5] ISAF: The International Security Assistance Force (in Afghanistan).

[6] NATO: The North Atlantic Treaty Organization.

10.5.2 Trajectories for International Engagement

The analysis in this chapter leads to a fundamental question: Can international engagement – diplomatic support, financial investment, capacity building, and discretionary funding – help provide a basis for ensuring water security in the Indus basin? Would such engagement be better directed toward integrating "Industan" or supporting efforts within each state?

The first question may presuppose a response to the second. Most national governments in the greater Indus region have, thus far, opted for domestic policies rather than regional ones; most international assistance actors lean more deliberately toward policies that lead to integration. This divergence is based on several factors: the accumulation of national interests (often supported by foreign assistance) that is based on state governance rather than regional efforts; the weakness of SAARC, the only regional political organization in South Asia; the state-led character and practices of foreign assistance; and, the institutional practices of water management in each state. Indeed, state actors tend to separate the management of the Indus – whether under the treaty (for India and Pakistan or separate from it, for the rest of the region) – from the effects of other policies and trends. For example, while international actors view climate change – a key factor in water security in a water-stressed region – as a global phenomenon, Indus basin states have generally considered phenomena associated with climate change as domestic and have designed their policies accordingly. Each state is a signatory to global agreements on climate change, but each continues to interpret its obligations independently. And, perhaps more importantly, the arenas in which water stress is most obvious – food and water security, social inequality, and differential political power – are deeply ingrained in national political psyches. The management of the Indus (and water more generally) has therefore been a domestic activity, occasionally belligerent and almost always protective.

For these reasons, international engagement that ranges far beyond the provision of financial and security assistance is likely to be crucial for the future security of the Indus basin states. If this comes to be true, it will be not only out of necessity but also because, as a first step, bilateral and multilateral donors will have to take seriously – certainly more seriously than before – the Paris Declaration precepts of neutrality, good governance, and cooperation. The latter is clearly the biggest challenge. While relationships among the Indus basin states have often been tension ridden on their own, they are also aggravated by decades of foreign policies that are supported (directly and indirectly) by foreign actors for reasons that have little to do with security and prosperity in the region. The contrast can be striking: it has been commonplace to note that the Indus Waters Treaty has defused tensions between India and Pakistan, but it is equally true that alliance politics have reinforced enmities at the same time.

One investment that international actors may find useful is to separate – and then explicitly and deliberately merge – the management of water security from the management of the region's bilateral and multilateral politics. International actors have expressed considerable interest in improving the management of the basin, a direction that could help substantially to reduce the risk of a range of water-related insecurities including food, disease, and livelihoods. This is the logic behind US intelligence analyses that identify the risks created by the combination of poor water management and high water pollution, inefficient agricultural practices, insufficient infrastructure, and climate variability (NIC 2012). The combined effects of these risks can threaten the political stability of each state and thus provoke tensions that threaten future regional peace.

The latter, potential, consequence is and should be the rationale for rethinking the external political support needed to sustain a relatively peaceful Indus basin region. All the foreign assistance projects in the world will not relieve the rigidities of relationships between India and Pakistan, or Pakistan and Afghanistan; history has already proved this. The imbalances in finances, population, and other resources between India and the region or China and the rest of the region (as well as the contests between India and China) will continue to drive a disputatious region unless and until all recognize the overriding importance of secure access to and management of water. In effect, the future of an imagined Industan may well rely on a future based on understanding water – an understanding based not on small-bore projects, but on a major rethinking of the region's external relationships.

Strategically, international engagement (bilateral and multilateral) may have the effect of changing minds. At the least, however, it can help to extend the time horizon for effective decision-making at the regional and national levels, cushion the risks for imaginative governance actors, and reduce the immediate likelihood of water-related conflict. If this becomes possible, then future engagements can help the Indus basin's residents to think bigger, and more fruitfully, about their futures.

References

Agrawal AK, Catalini C, Goldfarb A (2013) Some simple economics of crowdfunding, working paper 19133. National Bureau of Economic Research, Cambridge, Retrieved 22 Dec 2015 from: http://www.nber.org/papers/w19133

Ahmed V, Wahab MA (2012) Pakistan; nexus between aid and security. In: Kelegama S (ed) Foreign aid in South Asia: the emerging scenario. Sage Publishers, New Delhi

Ali I (2002) The historical lineages of poverty and exclusion in Pakistan. South Asia: J South Asian Stud 25(2):33–60

Arndt C, Jones S, Tarp F (2011) Aid effectiveness; opening the black box, UNU-WIDER working paper no. 2011/44. United Nations University World Institute for Development Economic Research, Helsinki

Briscoe J, Qamar U, Contijoch M, Amir P, Blackmore D (2005). Pakistan's water economy: running dry. Oxford University Press. Retrieved 22 Dec 2015 from: http://johnbriscoe.seas.harvard.edu/files/johnbriscoe/files/7._briscoe_-_qamar-_pakistans_water_economy-running_dry-_oxford_univ_press_2007.pdf

Burnside C, Dollar D (2000) Aid, policies, and growth. Am Econ Rev 90:847–868

Byrd WA (2016) What can be done to revive Afghanistan's economy? Special report 387. United States Institute of Peace, Washington, DC, Retrieved 18 Feb 2016 from http://www.usip.org/sites/default/files/SR387-What-Can-Be-Done-to-Revive-Afghanistans-Economy.pdf

Chandra V (2015) The unfinished war in Afghanistan 2001–2014. Pentagon Press, New Delhi
Chan S-J, Chung Y-H (2015) Trends and challenges of aid effectiveness: The rise of Asia. In: Cheng I-H, Chan S-J (eds) International education aid in developing Asia. Springer, Singapore
Cho C, Choi YC, Song YC (2014) Key development needs of South Asia and priority sectors of Korean ODA, world economic update. Korea Institute for International Economic Policy, Sejong-si
Cingano F (2014) Trends in income inequality and its impact on economic growth, OECD social, employment and migration working papers, no. 163. OECD Publishing, Paris
Cohen C (2007) A perilous course: U.S. strategy and assistance to Pakistan. Center for Strategic and International Studies, Washington, DC
Copestake J, Williams R (2014) Political-economy analysis, aid effectiveness and the art of development management. Dev Policy Rev 32(1):133–154
De Silva A (2012) Afghanistan; nexus between aid and security. In: Kelegama S (ed) Foreign aid in South Asia: the emerging scenario. Sage Publishers, New Delhi
Hout W (2002) Good governance and aid: Selectivity criteria in development assistance. Development and Change 33(3):511–528
Kaufmann D (2009) Aid effectiveness and governance: the good, the bad, and the ugly, development outreach. World Bank Institute, 26–29. Retrieved 23 Dec 2015, from http://siteresources.worldbank.org/EXTWBIGOVANTCOR/Resources/kaufmann-corrected.pdf
Kelegama S (ed) (2012) Foreign aid in South Asia: the emerging scenario. Sage Publishers, New Delhi
Klees SJ (2010) Aid, development, and education. Curr Issues Comp Educ 13(1):7–28
Lim AC-H (2015) The US, China and the AIIB: from zero-sum competition to win-win cooperation? – Analysis. Eurasia Rev. Retrieved 18 Feb 2016 from: http://www.eurasiareview.com/19042015-the-us-china-and-the-aiib-from-zero-sum-competition-to-win-win-cooperation-analysis/
Michaelowa K, Weber A (2007) Aid effectiveness in primary, secondary and tertiary education. UNESCO, Paris, Retrieved 23 Dec 2015, from: http://unesdoc.unesco.org/images/0015/001555/155559e.pdf
Morrow D, Carriere M (1999) The economic impacts of the 1998 sanctions on India and Pakistan. Nonproliferation Rev 6(4):1–16
Mosley P, Hudson J, Verschoor A (2004) Aid, poverty reduction and the 'new conditionality'. Econ J 114:217–243
Mustafa (2007) Social construction of hydropolitics: the geographical scales of water and security in the Indus basin. Geogr Rev 97(4):484–501
Neumayer E (2003) The determinants of aid allocation by regional multilateral development banks and United Nations agencies. Int Stud Q 47(1):101–122
NIC (2012) Global water security; intelligence community assessment, ICA 2012-08. National Intelligence Council (NIC), Washington, DC
OECD (2006). Harmonising donor practices for effective aid delivery, Volume 2, Budget support, sector wide approaches and capacity development in public financial management, Organization for Economic Cooperation and Development (OECD), Paris, France
Pronk JP (2001) Aid as a catalyst. Dev Change 32(4):611–629
Ravishankar N, Gubbins P, Cooley RJ, Leach-Kemon K, Michaud CM, Jamison DT, Murray CJL (2009) Financing of global health: tracking development assistance for health from 1990 to 2007. Lancet 373:3113–3124
Ravnborg HM, Larsen RK, Vilsen JL, Funder M (2013) Environmental governance and development cooperation achievements and challenges. Dan Inst Int Stud – DIIS Rep 2013:15
Schuster-Wallace CJ, Sandford R (2015) Water in the world we want. United Nations University Institute for Water, Environment and Health and United Nations Office for Sustainable Development, UNU-INWEH, Hamilton

Suhrke A (2006). The limits of statebuilding: the role of international assistance in Afghanistan, Paper presented at the International Studies Association annual meeting, San Diego, 21–24 March 2006. The Christian Michelsen Institute, Bergen

Tilak J (2014) South-South cooperation: India's programme of development assistance – nature, size and functioning. Asian Educ Dev Stud 3(1):58–75

Transparency International (2008) Global corruption report 2008; corruption in the water sector. Cambridge University Press, Cambridge, Retrieved 18 Feb 2016 from http://issuu.com/transparencyinternational/docs/global_corruption_report_2008?e=2496456/2011923

UN (2015) Transforming our world: the 2030 agenda for sustainable development, A/RES/70/1. United Nations, New York

UNDP (2006) Human development report 2006: beyond scarcity: power, poverty and the global water crisis. United Nations Development Programme (UNDP), New York

UNEP (2012) The UN-water status report on the application of integrated approaches to water resources management. United Nations Environment Programme (UNEP), Nairobi

Valdés A (2013) Agriculture trade and price policy in Pakistan, Policy Paper Series on Pakistan PK 17/12. World Bank, Washington, DC

Walker T, Sahin S, Saqib M, Mayer K (2014) Reforming electricity subsidies in Pakistan: measures to protect the poor, Policy Paper Series on Pakistan PK 24/12. World Bank, Washington, DC

World Bank (2013) Financing for development post-2015. World Bank Group, Washington, DC, Retrieved 18 Feb 2016 from: http://www.worldbank.org/content/dam/Worldbank/document/Poverty%20documents/WB-PREM%20financing-for-development-pub-10-11-13web.pdf

Yao Y (2011) Beijing consensus or Washington consensus: what explains China's economic success. World Bank, Washington, DC, Retrieved 23 Dec 2015 from https://openknowledge.worldbank.org/handle/10986/6098

Chapter 11
Conclusion

Robert G. Wirsing and Zafar Adeel

Abstract This chapter summarizes the contributing authors' main findings about water insecurity in the Indus basin. Recalling the reasons offered in Chap. 1 for the importance of the Indus basin to the world, it also includes the editors' reflections on the world's importance to the basin. By "the world" is meant those foreign powers that have had a serious and sustained strategic relationship with one or more of the basin's four riparian states, as well as the array of governmental and nongovernmental international and regional organizations and institutions that also have lengthy records of interaction with the countries sharing the basin. It also means the world order – the set of institutional arrangements, rules, and norms established internationally to promote stability, changes in which, in this highly interconnected and politically turbulent world, seem certain to impact upon water security in the countries sharing the Indus basin.

Keywords Integrated river basin management • Indus basin in the world order • Water resource cooperation

11.1 What Have the Authors Told Us?

As expected, the discourse presented in this volume on the Indus basin is dominated by the India-Pakistan relationship and the Indus Waters Treaty – as the sole legal, transboundary water management instrument at hand. However, the authors have deliberately pursued an expansion of that horizon to bring China and Afghanistan into sharp focus and essentially reconstitute the discussion to a larger geographical footprint that matches that of the Indus watershed. Despite paucity of published

R.G. Wirsing
Formerly, Georgetown University School of Foreign Service in Qatar, Doha, Qatar
e-mail: robertgw36@gmail.com

Z. Adeel (✉)
United Nations University, Institute for Water, Environment and Health, Hamilton, ON, Canada
e-mail: adeel_unu@yahoo.com

Z. Adeel, R.G. Wirsing (eds.), *Imagining Industan*, Water Security in a New World, DOI 10.1007/978-3-319-32845-4_11

literature and lack of legal instruments that relate to transboundary water management involving China and Afghanistan, the geopolitical and developmental analysis is presented as the basis for discussing a wider "Industan."

One of the most fundamental insights we may derive from the foregoing narratives is that the pre-Independence history of the Indus basin, specifically its history during the British colonial era, is one in which the basin's water resources were developed – and developed very extensively – in an almost entirely integrated or basin-wide manner. British water engineers built the basin's huge complex of irrigation canals, headworks, and dams unimpeded by the unyielding protocols of competing national sovereignties and fortified national boundaries. Irrigation requirements, not security imperatives, largely governed their planning. In short, the notion of Industan isn't entirely without precedent: indeed, the Indus basin's modern history includes a very lengthy period in which there existed essentially only one riparian state, British India, in whose manner of managing the basin's water resources may lie at least some guidelines for contemporary water planners.

About the post-Independence water resource history of the Indus basin, our authors were, of course, bound by the consequences of Partition of British India to tell us a very different story. The story they told us is necessarily well provisioned with conflicting national interests; deep-rooted hostility and distrust; domestic social, political, and institutional pathologies; and an exceptionally thin foundation of transboundary water agreements upon which to build. With an eye on these staple ingredients of the Indus basin, we summarize our authors' core findings in a threefold fashion, as follows.

First, any effort to foster cooperation over water resources in the Indus basin, whether at the local, regional, national, or international level, must labor against a myriad of diverse, complex, and frustrating handicaps. Conflict is deeply ingrained and resistant to change. To some degree, geography is destiny; and the basin's two lower riparian nations (India and Pakistan) are at the start unavoidably disadvantaged by their inherited geographic positions. These endow them with water dependency ratios that put them to a greater or lesser extent at the mercy of the basin's upper riparian nations – China in the case of India and Afghanistan, China, and India in the case of Pakistan. The power to dictate the terms of river resource extraction and use is spoken of as hydro-hegemony, and as Mark Zeitoun and Jeroen Warner observed in an influential article published in 2006, hydro-hegemony often lies at the heart of transboundary water conflicts. The advantages inherent in the upper riparian nation's geography-gifted control of water resources are a temptation that hydro-hegemonies find it very difficult to resist (Zeitoun and Warner 2006).

Geography is, of course, only one of many handicaps in the way of cooperation. Of those stemming from interstate hostility, the Indus basin has an unusually ample supply, including a number of notoriously intractable territorial disputes, lengthy histories of military and terrorist violence, memberships in rival strategic alliances, and intense competition over access to energy and other resources. Equally abundant, however, are those handicaps arising from circumstances in the riparian nations themselves. For instance, in Chap. 2 Majed Akhter argues that water scarcity in the Indus basin has far more to do with antiquated sociopolitical structures of inequality, rooted in what he terms "the agro-ecological transformations of the basin since the late 19th century," than with the decline in absolute or physical availability of per capita water supply. Quoting Saadia Toor, Akhter maintains that genuine land reform – "a radical program of land redistribution" – would do more to remedy Pakistan's increasing problem of water insecurity than building additional massive dams or, we might add, cursing India's hegemonic ambitions. Douglas Hill pursues a not dissimilar thematic trajectory in Chap. 8, which examines the potential for basin-wide management between India and Pakistan. Both Hill and Akhter reject the neo-Malthusian explanation of water scarcity which deems overpopulation as the primary source of the problem. Echoing Akhter, Hill suggests that

> the control of water in the Indus is embedded in the contestation between different parts of society and thus must be understood as the result of the interplay between bio-physical resources and broader political, social and cultural factors. In both countries, institutions of the state remain biased towards particular classes and regions, with the effect that water resource development in both India and Pakistan reinforces elite interests in dominant provinces at the expense of other groups.

Hill bemoans the "securitization" of water allocation debates in both countries. This tendency, he claims, in company with the dominant "technocratic approach to water resource management," has the effect of framing debates in terms of a narrowly defined "national interest" that encourages a focus on readily available technical fixes rather than on tackling insalubrious state-society relationships. Like Akhter, he argues that "the acceleration of large-scale dam construction in the Indus basin will intensify the negative consequences that we have seen in the past, both in terms of sustainable development and human security." Thus, Hill and Akhter trace the roots of water insecurity more to conditions *within* these countries than to conditions *between* them. "At the heart of the challenge of managing water across the Indus basin," writes Hill, "is the overcoming of poor governance on both sides of the border."

In his examination in Chap. 9 of the potential for basin-wide management between China and its co-riparian neighbors, India and Pakistan, Srikanth Kondapalli offers an approach accenting circumstances *between* the co-riparian neighbors. He draws our attention, in particular, to the power-balancing proclivities of Beijing's leaders, who, in designing plans for the country's future water security, place the pursuit of China's national interests far out in front of any concern for basin-wide management of the Indus. China, he says, has been investing very heavily since the late 1980s in the development of Pakistan's water resources; and with the announcement in April 2015 of a massive $46 billion aid package for Pakistan – more than half of which is earmarked for hydroelectric projects on several rivers (some of them in the politically highly sensitive area of disputed Kashmir) – Beijing, he claims, is clearly "upping the ante" in its relations with India. Guided by a Cold War-like logic, it steadily deepens and widens its water resource cooperation with Pakistan while settling for a very different – an "extremely slow and complicated" – process when it comes to forging river management agreements with India. For Kondapalli, China's increasingly strong embrace of its longtime "all weather" ally Pakistan, in company with its plans for unilateral exploitation of the rivers it shares with India, including the Indus, forecasts meager prospects for basin-wide management of the Indus.

The thesis advanced in Chap. 4 by Kristina Roic, Dustin Garrick, and Manzoor Qadir appears to present an interesting blend of domestic and geographic (riparian position) circumstances. Pakistan's unique and immense dependence on the Indus basin for its agricultural and other water requirements, they argue, when coupled with its lower riparian geographic placement, is the primary driver of Pakistan's water relations with India. "Being the lower riparian country," they observe, "Pakistan is hyper-sensitive to any activity in the Indus basin and tends to see its domestic water problems closely associated with India's actions upstream." There is thus a strong tendency, they say, for Pakistanis to blame India for their country's water woes.

Second, in spite of the evident difficulties confronting transboundary water resource cooperation, the catalogue of potential cooperative instruments is reasonably lengthy, and implementation of them is within the realm of possibility. In other words, there exists a fairly broad array of available and practical cooperative tools. Not every available cooperative instrument is discussed in this book. Of those that are, perhaps, the most obvious and commonly brought up is the legal instrument, specifically treaties or other sorts of more or less formal and binding river resource agreements. Many such international agreements and understandings, some bilateral and some multilateral, are explicitly identified in this book. The multilateral 1997 United Nations Convention on the Law of the Non-Navigational Uses of International Watercourses is brought into focus a number of times, but despite the fact that the 1997 Convention formally came into force in 2014 after 35 signatories ratified, none of the Indus basin riparian countries are signatories, a fact which signifies a lack of political interest in using that legal instrument. However, it is the bilateral 1960 Indus Waters Treaty (IWT) that is most often and most thoroughly discussed. Most of the chapters in this book at least mention the IWT, and Chaps. 1, 2, 5, and 8 treat of it in some detail – Bjørn-Oliver Magsig's Chap. 5 in considerable detail. Stating that the IWT mirrors a state-centric, zero-sum understanding of international relations, Magsig argues that the 1960 agreement is out of date, having few if any provisions that root governance of shared waters in contemporary common or collective security norms. He concedes that skepticism is warranted in regard to the near-term likelihood of regional or basin-wide management emerging in the Indus basin. He maintains, however, that progress toward a more resilient future for the basin can nevertheless still be made within the existing IWT framework – perhaps by amending it or, more simply, by breathing some life into the treaty's entirely neglected Article VII that urges the signatories to recognize their common interest in the Indus waters and "to co-operate, by mutual agreement, to the fullest possible extent."

Observing that climate change is certain to compound the already massive challenges confronting water managers in the Indus basin, David Michel in Chap. 6 argues that there is an abundance of opportunities for climate cooperation. All four of the co-riparian nations are up against complex and interwoven climate pressures and water-based security risks. He avers:

> Greater dialogue and coordination among the basin's diverse communities offers considerable scope for mitigating mutual threats and generating collective benefits. Collaborative approaches promoting joint monitoring and technical exchange, data sharing, and knowledge generation can help policymakers and the broader public better apprehend and assess the basin's complex climate and water challenges. Cooperative disaster preparedness and response can build regional resilience against future risks of worsening climate impacts. Common frameworks for collecting and disseminating policy lessons can enlarge the range of policy choices, scale up best practices, and chart potential pathways forward. The national policies of the riparian countries espouse such cooperative aims. Institutional structures exist to help enable collaborative strategies. Policymakers must enact them.

In view of the landmark Paris Agreement in December 2015, in which inter alia all the Indus basin riparian countries offered voluntary initiatives to combat climate change and adapt to new water/climate regimes, new opportunities for joint action have been created. How the countries respond to this new regime and whether the collective predicament of the region leads to a similarly collective political response remains to be seen.

In Chap. 7, Saleem Ali and Asim Zia reinforce Michel's argument about the acute vulnerability of the Indus basin co-riparians to intensifying environmental stress, in particular that arising from climate change, and about the resultant need for enlarged transboundary data sharing on environmental indicators. They appeal especially for greater support for science diplomacy, for cooperative research by Indian and Pakistani scientists in regard, for instance, to such shared problems as flood control, glacial melt, or monsoon variability. Of critical importance in facilitating science diplomacy, they point out, is to take advantage of such existing regional organizations as the South Asian Association for Regional Cooperation (SAARC) and the International Centre for Integrated Mountain Development (ICIMOD). "Transboundary data-sharing should be mainstreamed within South Asia," they conclude, "if the regional planning vision of 'Industan' is to reach fruition."

Zafar Adeel and Paula Newberg take the matter of organizational facilitation of water resource collaboration in the Indus basin a step higher to consider "how actors from outside the region can influence outcomes for overcoming water-related challenges." Such actors comprise the so-called international community, by which they mean "development banks, bilateral donor agencies, United Nations (UN) organization and agencies, internationally-driven nongovernmental organizations, and academic and research organizations …" The authors point out that a major and "truly transformative" shift occurred in the international development paradigm in late 2015 in the UN's adoption of the 2030 Agenda for Sustainable Development. Among the Agenda's extraordinarily comprehensive and ambitious Sustainable Development Goals (SDGs), water security remains a key to targeted economic, social, and environmental progress. While they concede that development assistance is no panacea for the Indus basin's manifold problems, Adeel and Newberg argue that international institutions can play a highly useful role in promoting collaboration, among other things by mobilizing financial assistance and by acting as trusted interlocutors or "honest brokers" for hydro-diplomacy in the Indus basin arena.

Third, the four co-riparian entities sharing the waters of the Indus basin will surely pay a huge price for noncooperation over these waters. They must move in the direction of cooperation, in other words, or suffer all manner of penalties – social, economic, political, and strategic. Virtually every chapter in this book, including the Introduction and this Conclusion, makes at least passing reference to the international and domestic costs of noncooperation over water resources in the Indus basin. This comes out most poignantly in Chap. 6, in which Michel warns that "a study of ten major Asian rivers concluded that the combination of mounting socio-economic stresses, increasing environmental strains, and persistent governance shortcomings render the Indus Basin most vulnerable of all to aggregate risk." He concedes that it is an open question whether the riparian nations are prepared to adopt a collective vision for managing the Indus basin's shared resources. In concluding, he suggests that the only certainty may be "that, if they fail to cooperate together in the face of climate change, then together they will bear its consequences."

In Chap. 3, Ashok Swain focuses entirely on these consequences, paying particular attention to India and Pakistan. The international costs include "intensification of traditional enmities, heightened distrust, persistence of territorial disputes, terrorism, and weakening of regional norms of cooperation." Besides these, a heavy price is also paid in the domestic context "in creating and aggravating the interprovincial and interethnic divisions within basin countries." How mistrust works to undermine the developmental goals of India and Pakistan is illustrated, for example, by the endlessly delayed Tulbul Navigation Project (aka Wular Project), launched by India initially in 1984. Designed to facilitate regular release of water from Wular Lake and year-round navigation on the Jhelum River, the project remains suspended yet today – for over 30 years – in the face of objections by Pakistan that the project is one of storage, not navigation, and that it is therefore in violation of the IWT. The stalling of this and a host of other Indian infrastructural projects on the rivers of the Indus system have, regardless of where the blame for delay lies, unquestionably done immense harm to India-Pakistan relations in general and to their water relationship in particular. "Recurring bilateral disputes over the water projects," Swain concludes, "have raised serious doubts about the possibility of developing the water resources of the river system in a mutually beneficial manner."

Skepticism about the prospects of even limited achievement of regionally integrated, basin-wide management of the Indus basin – of an imagined water-based Industan, in other words – is found in practically every chapter in this book. At least in the short term, a radical and swift transformation in the way water issues are dealt with in South Asia seems highly unlikely to all of its authors. This is not surprising given the political and strategic environment in which water planning has to be done. By the same token, all of this book's contributing authors have made abundantly clear their collective belief that much greater collective planning is essential, even unavoidable, if the Indus basin is to escape the likely disastrous consequences of continued failure to collaborate. There is agreement, in other words, that both the costs of failure and the difficulties of achieving something akin to Industan are equally daunting.

This brings us to consider, finally, why the world is important to the Indus basin, the reverse of the question we dealt with in the Introduction. We take this subject up because the future of the Indus basin, while obviously heavily contingent on the particular character of the four co-riparian nations, as well as on the priorities asserted and policies adopted by them, depends to a considerable extent on decisions taken elsewhere and the changes they bring about.

11.2 Why the World Is Important to the Indus Basin

In the Introduction to this volume, we offered six reasons why the Indus basin is important to the world. We conclude this volume now with our reflections on why the world is important to the basin. By "the world" we refer in particular to those foreign powers that have had a serious and sustained strategic relationship with one or more of the basin's four riparian states, as well as to the array of governmental and nongovernmental international and regional organizations and institutions, considered in the foregoing chapters, which also have lengthy records of interaction with the countries sharing the basin. But by "the world," we also mean to include the world order – the set of institutional arrangements, rules, and norms established internationally to promote stability, changes in which, in this highly interconnected and politically turbulent world, seem certain to impact upon water security in the countries sharing the Indus basin. Here we offer three reasons for the world's importance to the basin.

The *first* is that the Indus basin's four co-riparian states (Afghanistan, China, India, and Pakistan) have singly and collectively been a primary focus of great power rivalry for a good part of the twentieth century as well as the twenty first; and in the context of this basin, it has typically led to overall detrimental outcomes despite some nominal improvement. Since the withdrawal of the British from India in 1947, the world's great powers have entered into strategic alliances with, granted military and economic assistance to, sought to contain and even fought wars involving one or more of the four riparian states. All four of these states were heavily impacted by the Cold War between the Soviet Union and the United States. In its early decades, the Cold War turned Pakistan into America's "most allied ally" and China, following the routing of Kuomintang forces by the Communists in 1949, into a Soviet ally; and long before it ended with the collapse of the Soviet Union in 1990, the Cold War had turned New Delhi into one of Moscow's most valued Asian allies and a favored recipient of Soviet arms. It had also turned Afghanistan, following the coup d'etat that brought a communist regime to power in Kabul in 1978 and Soviet military occupation of the country soon thereafter, into a blood-drenched proxy battleground between the United States and the Soviet Union. The taking of power in Kabul by the Islamist Taliban in 1996 locked Afghanistan (and its neighbor Pakistan) into yet more decades of warfare. Lifted once again into prominence as a pressing global undertaking following the events of September 11, 2001, the war on terrorism swept up all four of the Indus riparian states, most especially the

Taliban-ruled Afghanistan, which suffered the immediate military consequences of American ire.

These military alliances and hostilities and the diplomatic maneuvering and intrigue that accompanied them invariably placed strategic interests far above water security or even economic and social development in the scale of national priorities – so far above, most of the time, that the basin's water security hardly registered at all as an international concern. While water security has in the last decade or so risen alongside other so-called "nontraditional" or "human" security issues to claim a larger share of international concern, it today still must compete for attention with "traditional," "hard," or power-driven security objectives. In this competition, water security still tends to fare rather poorly. There is a changing international narrative around water security that goes beyond the concept of "water wars" and starts to explore the consequences for society at large. A prime example of this thinking is the 2012 Intelligence Community Assessment by 24 US intelligence agencies which underlines the risk of internal collapse and societal disintegration in countries where coupled water and food insecurity is not managed adequately. Interestingly, the Indus basin figures prominently in this influential report and signifies its importance to the world.

Pakistan, in particular, has had great difficulty being dealt with as more than a pawn in great power strategic games – being recognized in its own right as a "hard" country, to draw upon Anatol Lieven's perceptive examination of Pakistan, unusually burdened with complexity and conflict, trapped in a myriad of befuddling enigmas, often misunderstood by others, and, above all, too important to ignore (Lieven 2012). Pakistan's fate at the hands of some American security analysts well illustrates Pakistan's identity crisis: used to being described in disparaging language in the American media, Pakistanis were not likely to have been surprised to see their country labeled "the ally from hell" in a 2011 article in the popular American monthly *The Atlantic* (Goldberg and Ambinder 2011). But they may not have been quite prepared for the scathing tongue-lashing Pakistan received in a 2015 issue of *Foreign Affairs*, America's leading international affairs journal of opinion. In this essay, written by the prominent American scholars C. Christine Fair and Sumit Ganguly, the need to discard entirely Washington's decades-old security alliance with Pakistan – astonishingly, even to reclassify Pakistan as an enemy – is unequivocally urged. All past American efforts to persuade Pakistan to change its ostensibly perverse and malicious behavior have mainly failed, the authors point out, and there is no longer any convergence of interests between Washington and Islamabad. "After six decades of policy predicated on Pakistani blackmail," they write, "it should be possible to achieve U.S. interests with a different approach. A strategy of containment is the United States' best option. Above all, U.S. relations with Pakistan should be premised on the understanding that Pakistan is a hostile state, rather than an ally or a partner" (Fair and Ganguly 2015).

Now their previous writings make it certain that Fair and Ganguly would never be mistaken as turncoats – once Pakistan sympathizers by the force of events turned critics. On the contrary, both have for long been consistently critical of what they understand to be Pakistan's role in the region and especially of what they characterize as its army leadership's relentless, counterproductive, and self-serving rendering of India as Pakistan's mortal enemy. Unfortunately, their shortsighted, bullying, and uncompromising assault on Pakistan does not merely remind us of the harsh but realistic observation attributed to Lord Palmerston that "nations have no permanent friends or allies, they only have permanent interests." Indeed, in the unlikely event their palpably irresponsible advice is taken seriously in Washington, DC, the called-for radical turnaround in American strategic commitments would obviously deal a heavy blow to Pakistan's standing in the world, add immeasurably to its threat calculus, drive it still more closely into the arms of China, and perhaps even inspire reckless military action. None of that would seem likely to foster more congenial relations with India, a long hoped-for development which, in our judgment, ought to be among the key objectives of American strategic planning for the region. Were Pakistan actually to be unceremoniously "cut loose," those whose already difficult task it is to lift water security up the ladder of international priorities in the Indus basin obviously would then be confronted with an insurmountably troublesome conundrum: how to put customarily adversarial co-riparian nations on the road to transboundary water resource cooperation when a drastic strategic about-face was forcefully driving them apart.

A *second reason* why the world is important to the Indus basin is that the world order imposed unilaterally by the United States at Bretton Woods near the end of World War II appears to many observers to be losing ground – to be weakening, in other words, in the face of challenges to America's hegemonic position coming from such rising powers as China and India. Obviously, with these two revisionist states ranking as the largest and most powerful of the Indus basin riparian nations, if, when, and how the world order changes will impact the Indus basin hugely and in many ways, not least in helping to determine whether the countries sharing the basin take the cooperative or conflictive road in regard to transboundary water resources.

Speculation on the likely shape of a new world order is now a popular pastime; and analysts are routinely served with a feast of momentous questions. In response to one of them – what are the implications for the world order of China's rise? – a near consensus seems to have arisen that global power is shifting perceptibly to the East and that China's rise is necessarily accompanied by America's decline. Further, given the developing instability in the global balance of power, the world should ready itself even for the unthinkable – the outbreak of a war between the United States and China. On this subject some published works have obviously been driven by sensationalism. But not all of them are guilty of this. In *The Improbable War: China, the United States and the Logic of Great Power Conflict*, for example, Christopher Coker, professor of International Relations at the London School of Economics, presents an eminently well-reasoned and sobering argument that war between these two states, while by no means inevitable, is definitely possible. The book ends with Coker's warning that "it is of vital importance that the possibility of a conflict between China and the United States continues to be discussed because the slippage, flux and confusion of the tide of history can lead to many different outcomes, and it is important to remember that the prevailing complacency regarding the obsolescence of great power war contributed to the outbreak of war in 1914. If the United States and China continue to convince themselves that war is too 'improbable' to take seriously, it is not only they but the rest of the world that may ultimately pay the price" (Coker 2015).

Of course, not everyone goes along with these findings. In his most recent book *Is the American Century Over?*, for instance, Joseph S. Nye, Jr., one of America's most seasoned international affair scholars, maintains that the United States is a far more durable great power than many suppose and that the often claimed irreversibility of its decline, as though some inexorable historical forces were behind it, amounts to a basic misreading of how nations manage their power and how the world order is actually determined (Nye 2015). Along the same lines, David Shambaugh, commenting on China's capacity to challenge America's hegemonic position, concedes in *China Goes Global: The Partial Power* that China has indeed made remarkable economic progress and that in some categories of national power, including conventional military and unconventional cyber warfare capability, it is clearly catching up with the United States. But Shambaugh throughout the book relentlessly pushes his thesis "that China has a very long way to go before it becomes – if it *ever* becomes – a true global power. And it will never 'rule the world'" (Shambaugh 2014). And in the arrestingly provocative, perceptive, and deeply troubling book by Peter Zeihan, *The Accidental Superpower: The Next Generation of American Preeminence and the Coming Global Disorder*, the argument is made that it is China, not the United States, that is going downhill and that China is in no position to lay claim to the title of world hegemon (Zeihan 2014).

Zeihan's most arresting argument – one with particular pertinence for the countries sharing the Indus basin – is that the United States, protected in its economically increasingly autarkic North American fortress by the world's most favorable geography, demography, and resource endowment, is bound to lose interest in acting as the world's policeman, a terrifically costly duty it took on after World War II in return for the world's acquiescence in the Bretton Woods system of free trade. His depiction of "the descent" into international disorder following upon America's withdrawal from its free trade-policing duties, which he expects to happen in the next few decades, is worthy of the descent to the pit of Hell in Dante's *Inferno*. "The global financial wave," he observes,

> will crest at some point between 2020 and 2024…. Between 2020 and 2024, 13 of the world's top 25 economies will be in the ranks of the financially distressed.... Aging demographies will sharply and suddenly contract credit availability to a level that has not been witnessed since the 1970s – in the best case. Interest and mortgage rates will climb into the teens in the developed world, and higher in the developing world. … The pace of technological change will screech to a halt…. Governments the world over will have to make ever more difficult decisions…. The international economy will spasm and contract…. Consumption of both raw commodities and finished goods will plummet…. Everywhere, American power will be overwhelming by its absence. For countries like China, which are dependent upon exports to the American market, the pain will be direct and permanent. Others – Central Europe comes to mind [and Afghanistan, Pakistan, and India perhaps?] – will suffer from the withdrawal of American military support (ibid. pp. 148–150).

If this speculation weren't already discomfiting enough, Zeihan goes on to say that "*that* is the positive scenario, because it assumes that everyone gets along. It is far more likely that they won't." Why won't they? Because, he says,

> Countries far removed from supplies of food, energy, and/or the basic matrix of inputs that make the industrialized world possible will face the stark choice of either throwing themselves at the mercy of superior local powers or throwing what force they can muster at the resource providers. In their desperation, many will realize that American disinterest in the world means that American security guarantees are unlikely to be honored. Competitions held in check for the better part of a century will return. Wars of opportunism will come back into fashion. History will restart. Areas that we have come to think of as calm will seethe as countries struggle for resources, capital, and markets. For countries unable to secure supplies (regardless of means), there is a more than minor possibility that they will simply fall out of the modern world altogether. (ibid. pp. 150–151)

Zeihan is far from being the only Jeremiah-like global-order prophet of doom among international relation scholars. Robert Kaplan and Randall L. Schweller, for instance, both offer equally apocalyptic visions of the future (Kaplan 2001; Schweller 2014). One hopes that they have it all wrong. But even if they prove to be only partially right, the implications of the changes they see coming leave little room for complacency about resolving problems of water scarcity and, in particular, about resolving conflicts over transboundary water resources.

The *third and final reason* we offer why the world is important to the Indus basin emerges directly from the second. It is that there is an extremely urgent need for the community of nations, however organized and institutionalized, to labor still more earnestly to lay a foundation for the coming world order that facilitates the steady growth of rules and norms (and of the institutions that foster and protect them) that encourage cooperation in the management of transboundary water resources. This is, of course, a very tall order – especially in view of the apocalyptic visions noted above. Happily, at least some of the foundation has already been laid.

The United Nations-sponsored 1997 Convention on the Law of Non-Navigational Uses of International Watercourses is clearly one important step in the laying of this foundation. On 19 May 2014, the Convention entered into force for the ratifying states 90 days following the deposit by Vietnam with the Secretary-General of the United Nations of the 35th instrument of ratification, acceptance, approval, or accession. There have now been (as of 7 October 2015), with ratification on 6 January 2015 by the State of Palestine, a total of 36 ratifications. The fact that it took the Convention over 17 years after its adoption by the United Nations General Assembly to garner ratifications from a relatively small number of United Nations member states highlights the worldwide resistance by national governments to become party to a legal instrument that may ostensibly have an adverse impact on national sovereignty, perceived strategic interests, existing transboundary agreements, and strength of relationship with upstream/downstream riparian countries (Salman 2015).

It takes little more than a glance at the list of ratifying states to realize that the step taken in 1997 was, in fact, relatively modest, indeed in some respects quite discouraging. Among the ratifying states, there is not a single East or South Asian state, and there is only one Central Asian and one Southeast Asian state. There is also not a single ratifying state among them from the Western Hemisphere. The largest and most powerful Asian state (China) was one of three (Turkey and Burundi are the other two) to vote against adoption of the Convention; five other Asian states, including three of those sharing the Indus basin (India, Pakistan, Afghanistan), either abstained or were absent. Six more Asian states did vote for the Convention, but of these six, only two (Uzbekistan and Vietnam) have ratified it. Neither of Asia's most powerful upper riparian states, China and India, threw its support to the Convention.[1]

Authoritative analyses of the Convention since its adoption have made clear that it suffers not only from the paucity of ratifications but also from a variety of content weaknesses. Among the more important is that its central normative pillars – the obligation of states sharing transboundary rivers not to cause harm to their co-riparian neighbors and to make use of shared waters in an equitable and reasonable manner – are vague in meaning, subject to rival interpretations, potentially contradictory, and a feast for lawyers. Also often brought up as a weakness is that dispute settlement provisions are nonbinding (Salman 2007, 2015).

While the number of international legal experts who consider the Convention of little or no value is fairly substantial, it seems that a majority share the opinion that it is likely to carry significant weight in cases of transboundary water resource conflict even had it failed to get the required number of ratifications. According to one careful evaluation of the Convention, its virtues include the following:

- Having been developed by leading experts assembled by the International Law Commission, it has the stamp of scientific authority and is thus not easy to ignore.
- The equitable use and no harm principles are mainly codifications of already existing norms.
- Its authority gains from the fact that virtually every interested state had opportunity to participate in the negotiations.
- The Convention was adopted in the general assembly by a large number of countries (103 United Nations member states voted in favor).
- Finally, the Convention influenced the drafting of language of specific regional agreements and that influence extends even retroactively to the 1996 watercourse agreement on the Ganges River concluded by Bangladesh and India (Schroeder-Wildberg 2002).

Regarding the Indus basin itself, the authors in this book have touched upon a number of themes which may be counted among positive normative achievements in line with and likely inspired by legal advances at the global level. Included, of course, would be the Indus Waters Treaty, which, in spite of its substantial inadequacies highlighted in this book, does contain impressive conflict prevention provisions that have displayed considerable robustness over the years. Perhaps the most notable achievement of these provisions are the awards delivered by the Hague's

[1] The regional breakdown of ratifications is the following: European (16), African (12), Middle Eastern (6), Central Asian (1), and Southeast Asian (1).

Permanent Court of Arbitration (PCA) in the case brought by Pakistan in 2010 in regard to India's Kishenganga Hydroelectric Project on the Jhelum River in Indian-controlled Jammu and Kashmir.

The sequence of PCA awards, announced on 18 February and 20 December 2013 (PCA *Partial Award*; PCA *Final Award*; PCA *Decision* 2013), obviously does not carry enough weight to overcome overnight the enormous distrust and clashing interests chronic in India-Pakistan relations. However, the Court's rulings not only on river diversion and dam design features of the Indian project but also – and quite innovatively – on environmental sustainability supply a much-needed normative precedent for the entire Indus basin. As Upadhyay and Chatterjee observe in the *Indian Journal of Arbitration Law*, "the Kishenganga dispute serves as a valuable reminder of the potential of pacific dispute settlement in resolving thorny disputes in tense situations. For India and Pakistan, the award attempts to exemplify the co-operative spirit that underlies the IWT and to strike a fine balance between the competing rights of the two states. More broadly, the award represents an interpretive approach towards technical treaties informed by contemporary international law principles relating to environmental protection and sustainable development." They then conclude that "the rendering of the Kishenganga partial award by the PCA clearly goes on to exemplify that regardless of the high stakes involved in the dispute, [or] the political background of the aggrieved parties, the Court has carved out a well-balanced solution setting a precedential example for other such river water disputes to follow" (Upadhyay and Chatterjee 2015).

These PCA awards should not be deprecated as isolated cases of arbitration that will soon pass into the dustbin of history. Awards like them are in fact becoming a fairly regular feature of bilateral dispute resolution around the world and, interestingly, also in the South Asian region. On 14 March 2012, for instance, the International Tribunal for the Law of the Sea (ITLOS), an arbitral body created by mandate of the Third United Nations Conference on the Law of the Sea in 1982, rendered its judgment delimiting the territorial sea of Bangladesh and Myanmar as well as each state's share of the Exclusive Economic Zone (EEZ) and the continental shelf (ITLOS 2012). Accepted by both countries as a fair decision, the judgment has been praised not only for the swiftness, transparency, and legal innovativeness of the proceedings that produced it but also as "perhaps evidence of a positive new trend in international dispute resolution as a whole" (Riesenberg 2012). The subject of arbitration was far from trivial: arbitration had been resorted to following decades of inconclusive negotiations, enormously complicated by the discovery of oil and gas in the disputed waters.

A second maritime boundary dispute to be resolved in the same fashion was that between India and Bangladesh. Ending a 40-year-long quarrel bearing many similarities with that between Bangladesh and Myanmar, an arbitral tribunal set up under the Hague's PCA rendered a judgment on 7 July 2014, which, while awarding Bangladesh a larger share of the disputed maritime area (76 %), ruled in favor of India on a number of issues – enough so that both sides welcomed the verdict (Arbitral Tribunal, *Maritime Boundary Arbitration* 2014). Again, the subject of arbitration was of huge importance to both countries: with their maritime boundaries now clearly delimited and freed of legal uncertainties, a vast maritime area, totaling almost 407,000 km^2, was open for both countries to exploit potentially vast offshore energy resources. Important to note here is that the three neighboring (but not consistently neighborly) countries involved (India, Bangladesh, Myanmar) displayed commendable willingness to submit to arbitral proceedings that were not subject to appeal, and they also demonstrated admirable willingness to accept verdicts that granted none of them all that they wanted.

11.3 A Peek into the Crystal Ball

There is considerable interest in understanding and visualizing scenarios and potential outcomes for the Indus basin as a whole. Based on the discourse presented in this concluding chapter, definitive answers to complex questions lying at the intersection of water security, regional hegemonic interests, interplay of external players, aspirations of sustainable development, and changing climate/water patterns are not readily available. However, through the arguments presented in their respective chapters and even more candid dialogue during an authors' workshop organized in Hamilton in September 2015, the collective wisdom of the authors points to at least partially favorable outcomes. We tend to lean in that direction also because of two reasons: First, the collective weight of positive arguments – particularly political and economic – in favor of a more integrated, holistic management of water resources could tip the balance in that direction. Second, the macroscale changes to management require political support at the highest level, and we have observed the ability of leadership in all four riparian countries to move swiftly to make grand decisions when the outcome is perceived to be politically favorable; the China-Pakistan collaboration on development of a $46 billion economic corridor is the clearest example of such decision-making.

We also do not want to leave the impression here that the globally generated rules and norms required to encourage cooperation in the conduct of bilateral relations have now been fully internalized in the South Asian region and can be counted upon to govern negotiations among the countries of the Indus basin about the management of transboundary water resources. Nothing of the kind is presently visible. We do, however, want to reaffirm our belief that it is precisely the adoption of such rules and norms – and the institutionalization and routinization of their practical application – that is needed to forestall (or at least make less likely) the emergence of the Hobbesian *bellum omnium contra omnes* that some think is the fate of the present century.

At the end of the day, the fate of an integrated Industan rests with the riparian countries, their peoples, and respective governments. The arguments, data, and information presented in this volume – particularly that drawn from other, similar basins – adds to the overall evidence base for the political decision-making and normative changes.

References

Arbitral Tribunal of the Permanent Court of Arbitration (2015) In the matter of the bay of Bengal Maritime Boundary Arbitration (the People's Republic of Bangladesh and the Republic of India), 7 Jul 2014, at: http://www.pca-cpa.org/showfile.asp?fil_id. Retrieved 27 Apr 2015

Coker C (2015) The improbable war: China, the United States & the logic of great power conflict. Hurst & Company, London, p 181

Fair CC, Ganguly S (2015) An unworthy ally: time for Washington to cut Pakistan loose. Foreign Aff 94(5):160–170

Goldberg J, Ambinder M (2011) The Ally from Hell. The Atlantic, at: http://www.theatlantic.com/magazine/print/2011/12/the-ally-from-hell/8730/. Retrieved 3 Oct 2015

International Tribunal for the Law of the Sea (ITLOS), Case No. 16, Dispute Concerning Delimitation of the Maritime Boundary between Bangladesh and Myanmar in the Bay of Bengal, 14 March 2012, at: http://www.itlos.org. Retrieved 27 Apr 2015

Kaplan R (2001) The coming anarchy: shattering the dreams of the post cold war. Random House, New York

Lieven A (2012) Pakistan: a hard country. Public Affairs, New York

Nye JS Jr (2015) Is the American century over? Polity, Boston

Permanent Court of Arbitration, Decision on India's Request for Clarification or Interpretation in the Matter of the Indus Waters Kishenganga Arbitration, 20 December 2013, at: http://www.pca-cpa.org/PK-IN%20Decision%20on%2. Retrieved 8 Oct 2015

Permanent Court of Arbitration, Final Award in the Matter of the Indus Waters Kishenganga Arbitration, 20 December 2013, at: http://www.pca-cpa.org/PK-IN%20Final%20Award,%2. Retrieved 8 Oct 2015

Permanent Court of Arbitration, Partial Award in the Matter of the Indus Waters Kishenganga Arbitration, 18 February 2013, at: http://www.pca-cpa.org/PK-IN%2020130218%20Partie. Retrieved 8 Oct 2015

Riesenberg DP (2012) Introductory Note to the International Tribunal for the Law of the Sea Delimitation of the Maritime Boundary between Bangladesh and Myanmar in the Bay of Bengal (Bangladesh/Myanmar), at: http://www.jstor.org/discover/10.5305/intelegamate.5. Retrieved 27 Apr 2015

Roughton D et al (2015) UNCLOS Annex VII Tribunal Decides Bangladesh-India Maritime Boundary Dispute at: http://www.hsfnotes.com/arbitration/2014/07/15/unclos-annex-vii-tribunal-decides-bangladesh-india-maritime-boundary-dispute. Retrieved 27 Apr 2015

Salman SMA (2007) The United Nations watercourses convention ten years later: why has its entry into force proven difficult? Water Int 32(1):1–15, http://www.unwatercoursesconvention.org/images/2t, Retrieved 7 Oct 2015

Salman SMA (2015) Entry into force of the UN watercourses convention: why should it matter? Int J Water Resour Dev 31(1):4–16

Schroeder-Wildberg, E (2002) The 1997 International Watercourses Convention – Background and Negotiations, Working Paper on Management in Environmental Planning Number 004, at: http://www.landschaftsoekonomie.tu-berlin. Retrieved 7 Oct 2015

Schweller, R L (2014) Maxwell's Demon and the Golden Apple: Global Discord in the New Millenium. Johns Hopkins University Press, Baltimore

Shambaugh D (2014) China goes global: the partial power. Oxford University Press, Cambridge, p 6

Upadhyay A, Chatterjee T (Forthcoming, Dec 2015) The Kishenganga hydro-electric project arbitration dispute – partial award (Pakistan v. India): an analysis. Indian J Arbitr Law III(2), at: http://ijal.in/sites/default/files/Tamojit.pdf. Retrieved 8 Oct 2015

Zeihan P (2014) The accidental superpower: the next generation of American preeminence and the coming global disorder. Twelve Books, New York

Zeitoun M, Warner J (2006) Hydro-hegemony: a framework for analysis of trans-boundary water conflicts. Water Policy 8(5):435–460

Index

Z. Adeel, R.G. Wirsing (eds.), *Imagining Industan*, Water Security in a New World, DOI 10.1007/978-3-319-32845-4

MIX
Papier aus verantwortungsvollen Quellen
Paper from responsible sources
FSC® C105338

If you have any concerns about our products,
you can contact us on
ProductSafety@springernature.com

In case Publisher is established outside the EU,
the EU authorized representative is:
Springer Nature Customer Service Center GmbH
Europaplatz 3, 69115 Heidelberg, Germany

Printed by Libri Plureos GmbH
in Hamburg, Germany